Historical & Cultural Astronomy

More information about this series at http://www.springer.com/series/15156

Giulio Magli • Antonio César González-García •
Juan Belmonte Aviles • Elio Antonello
Editors

Archaeoastronomy in the Roman World

Springer

Editors
Giulio Magli
Department of Mathematics
Politecnico di Milano
Milan, Italy

Antonio César González-García
Instituto de Ciencias del Patrimonio
Incipit-CSIC
Santiago de Compostela, Spain

Juan Belmonte Aviles
Instituto de Astrofisica de Canarias
La Laguna, Spain

Universidad de La Laguna
La Laguna, Spain

Elio Antonello
INAF-Osservatorio Astronomico di Brera
Merate, Italy

ISSN 2509-310X ISSN 2509-3118 (electronic)
Historical & Cultural Astronomy
ISBN 978-3-319-97006-6 ISBN 978-3-319-97007-3 (eBook)
https://doi.org/10.1007/978-3-319-97007-3

Library of Congress Control Number: 2019930545

This Springer imprint is published by the registered company Springer Nature Switzerland AG
The registered company address is: Gewerbestrasse 11, 6330 Cham, Switzerland

Foreword

Past societies—not only during prehistory—made gods of celestial bodies and phenomena, including them in their mythological narratives and linking them not only to atmospheric events and seasonal cycles but also to important social institutions and the destiny of men.[1] Early on in the history of archaeology, some researchers, especially astronomers but also anthropologists and prehistorians, began suggesting that some archaeological remains (especially prehistoric ones) could be measured and examined in order to explore potential alignments with celestial phenomena. This, they claimed, would improve our understanding of the symbolic universe of the groups and societies that built these monuments. In time, this developed into an interdisciplinary subject of study, half way between astronomy and the social sciences.[2] Different terms have been put forward to refer to the discipline, including archaeoastronomy (E. Ch. Baity; M. Hoskin), astroarchaeology (Hawkins 1973), the history of astronomy and ethnoastronomy; more recently, S. Iwaniszewski (1997) and C. Ruggles (1999, 2001) proposed an all-embracing term, cultural astronomy (which can accommodate both the history of astronomy and ethnoastronomical traditions). This discipline examines how, throughout history, humans have oriented themselves in time and space through the observation of celestial bodies (Belmonte 2009: 58).

Within the field of cultural astronomy (Iwaniszewski 2009: 30), archaeoastronomy has been characterised by a lasting and intense debate around the discipline's very definition and methodological orientation. In recent decades, some degree of

[1]See, for instance, Lehoux (2007); Silva and Campion (2015). From prehistory until today (Chamberlain et al. 2005), groups and societies developed relatively advanced astronomical and calendrical knowledge, which served a variety of purposes. Calendars and programmed agricultural activities were made possible by the study of the position of celestial bodies (Antonello 2012).

[2]In general, this discipline analyses the alignment of monuments and buildings (menhirs, tombs, temples, etc.) and celestial phenomena (e.g. solar and lunar dawns and sunsets). The discipline also examines the iconographic representation of celestial bodies and phenomena and the reconstruction of astronomical events using the data gathered by modern astronomy (Krupp 1989, 1991).

agreement has been reached about the difficult question of definition. The discipline is thus described as an approach to the astronomical knowledge and beliefs of past societies (Belmonte 2000: 14) or, more restrictedly, to the astronomical practices of prehistoric societies (Esteban 2003a: 309). That is, the discipline examines astronomical, archaeological, historical, ethnographic and anthropological data in order to investigate the interaction of men and the cosmos from prehistory to the present day (Cerdeño et al. 2006: 14).

Methodological debates have also played a central role in the discipline during the final decades of the twentieth century. These debates revolved around the need to conciliate the methods followed by physicists and mathematicians (who can analyse the motion and position of celestial bodies) and those adopted by archaeologists, historians, ethnographers, etc., who can examine the cultural patterns of past societies and are in a position to ask the right questions (Silva and Campion 2015).[3] As such, the relationship between archaeology and astronomy relies on the ability of the former to provide data (gathered by means of methodologically precise astronomical calculations), the cultural analysis of which by the latter can contribute to the interpretation of the material record of past societies.[4] In addition to this, archaeology as a historical discipline has notably expanded its chronological boundaries over time, and multiple specialised period-specific archaeologies exist (Gutiérrez 1997: 25–88); it is worth emphasising that the data collected by physicists about a given period must be interpreted by archaeologists who specialise in that period.

Archaeoastronomy has a long tradition in some European countries, where there was an early interest in the interaction between past societies and the cosmos (Morellato 2011). The origins of the discipline go back to the late seventeenth century,[5] when a number of British antiquarians began to interpret megaliths as

---

[3]However, the twofold nature of the discipline has resulted in some degree of theoretical and methodological confusion and in the emergence of epistemologies in conflict (Iwaniszeski 1994: 5; 2003). For this reason, the interdisciplinary cooperation of physicists and archaeologists is essential for the appropriate data to be collected (Cerdeño and Rodríguez-Calderot 2009: 282–284; Esteban 2009: 69–77) and given a sound cultural interpretation. Some have even argued for the need to create a new professional category, that of the archaeoastronomer, in which the skills of both fields can meet (Antonello 2012; Belmonte 2009: 59); this is not impossible, but being proficient in the skills of both disciplines looks like a rather difficult task. Recent projects reflect the complexity of the issue, such as the *Journal of Skyscape Archaeology* (the publication of which began in 2015) (https://journals.equinoxpub.com/index.php/JSA), which aims to be a platform for the analysis of the archaeological record from the point of view of celestial phenomena, analysing the relationship between material culture, cosmos and society throughout history. The journal promotes a multidisciplinary perspective and encourages archaeologists to expand their horizons and include the sky in their cultural landscapes, while compelling archaeoastronomers to focus their study on the cultural interpretation of the material record.

[4]That is, archaeoastronomy would essentially be a technical discipline, a form of archaeometry, that is, a methodology which provides data to archaeologists for their subsequent interpretation (Cerdeño and Rodríguez-Calderot 2009: 279–286).

[5]During this period, archaeology was limited to the work of a number of antiquarians and their sponsors, while the earliest academies and museums began to open their doors; during this period, excavations were initiated in Pompeii and Herculaneum, and travellers began reporting discoveries

'astronomical observatories'.[6] Although the term 'archaeoastronomy' was used for the first time by Elizabeth Chesley Baity in 1973, the roots of the discipline are still a matter of debate. While such important physicists as Heinrich Nissen and Norman Lockyer (active in the late nineteenth and early twentieth centuries) could be considered as the fathers of modern archaeoastronomy, most agree that the discipline truly hatched in the United Kingdom, with Alexander Thom, a Scottish engineer who worked in England, especially at Stonehenge, from the interwar period to the 1970s.[7]

The 1980s witnessed the consolidation and growth of the discipline and the dispelling of numerous myths concerning various prehistoric 'observatories'. The discipline expanded into new geographical areas (the Balearic Islands, Sardinia, the Iberian Peninsula, America, Africa, etc.) and incorporated scholars from multiple countries (not only English-speaking), vindicating its multidisciplinary nature and demanding a space in the academic universe. In this context, the Leicester archaeologist Clive Ruggles, who re-examined Thom's data and arguments, and the Cambridge mathematician Michael Hoskin pushed for the creation, within the framework of the International Astronomical Union (IAU), of *The 'Oxford' International Symposia on Archaeoastronomy* in 1981.[8] Since its inception, this body has endeavoured to unify scientific and archaeological data and interpretations.

In the 1990s, the discipline underwent a phase of unprecedented growth, with the inclusion of yet more geographical regions and cultural horizons and with the publication of the earliest regional syntheses (Belmonte 1994; Romano 1992). This phase also witnessed the end of 'monumentalist' approaches,[9] and the

---

in the Eastern Mediterranean; the earliest repertoires of antiquities were also published during this period, and J.J. Winckelmann outlined the principles of archaeological science as the history of Greek art; prehistory was barely defined as a discipline (Bianchi Bandinelli 1992[2]).

[6]John Aubrey (in 1678) and Henry Chauncy (in 1700) analysed some of the astronomical principles that governed the orientation of medieval Christian churches, while in 1740 the architect J. Wood and the antiquarian William Stukeley studied the astronomical orientation of the megalithic assemblages of Stonehenge's sarsen circle and Callanish, among others, presenting the idea of British (and later European) megaliths as astronomical observatories. Their ideas would remain virtually unchallenged until the 1980s. On the other hand, in the late nineteenth century, the astronomers Richard Proctor and Charles Piazzi Smyth examined the astronomical orientation of the pyramids of Giza, in Egypt, inaugurating the archaeoastronomical study of the major pyramid-building cultures, such as the Egyptian and the Maya (Aveni 1991; Bauer and Dearborn 1998; Galindo 1994; Šprajc 2001).

[7]The early scientific phase of the discipline, which focused on the measurement of astronomical orientations rather than on historical and cultural interpretation, led to the creation of the *Journal for the History of Astronomy* (1970) and later of its supplement, *Archeoastronomy* (1979). Although his work has been subject to a profound revision, Thom's influence persists, and his statistical analysis methodology remains part of the basic toolkit of the archaeoastronomer (Thom 1954: 396–404; 1967; 1984: 129–148).

[8]This body has celebrated a total of 11 symposia to date. The twelfth one is scheduled for celebration in La Plata, Argentina, in 2020; see https://www3.archaeoastronomy.org/index.php/oxford-conferences

[9]An approach that, to some extent, has also hampered Classical Graeco-Roman archaeology until recent times.

consolidation of stable avenues of cooperation between astronomers and archaeologists, as illustrated by the Stonehenge-centred project directed and published in 1997 by B. Cunliffe and C. Renfrew (1997). Another important milestone was the foundation of the *Société Européenne pour l'Astronomie dans la Culture* (SEAC) (Strasburg 1993), by the astronomer C. Jaschek.[10] This was followed in 1996 by the inception of the *International Society for Archeoastronomy and Astronomy in Culture* (ISAAC), created in the United States with the aim of developing the academic presence of archaeoastronomy and ethnoastronomy;[11] the *Sociedad Interamericana de Astronomía en la Cultura* (SIAC)[12] was founded in Santiago de Chile in 2003.[13] In recent decades, these three associations have worked ceaselessly for the promotion of archaeoastronomical studies (Belmonte 2016: 93–101).[14]

Archaeoastronomy is currently a mature discipline practised worldwide, with a place in the academic arena,[15] awake to theoretical and methodological concerns, and capable of producing rigorous results. This maturity is also reflected in the publication of synthetic works such as the monumental *Handbook of Archaeoastronomy and Ethnoastronomy.*[16] Another important outcome of the growth of the discipline is the cataloguing of 'astronomical' sites, their potential recognition as

[10]It has been pointed out that the work carried out within the framework of this body focuses excessively on technical astrophysical matters and lacks archaeological interpretation; see: http://www.archeoastronomy.org/

[11]https://www3.archaeoastronomy.org/

[12]Constituted by professionals working in the astronomical and cultural fields, from the point of view of archaeoastronomy, ethnoastronomy and the history of astronomy; see: http://eacultural.fcaglp.unlp.edu.ar/

[13]Within the framework of the *Simposio de Etno y Arqueoastronomía del Congreso Internacional de Americanistas.*

[14]This work includes promoting the field in universities; the development of interdisciplinary cultural astronomy studies; the creation of links between international, regional and national experts; and the organisation of symposia, workshops and field schools, which have channelled most of the scientific activity of the discipline and have become the main arenas for debate and the presentation of results. Since 1993, the SEAC has organised 25 conferences (apart from the foundational conference, celebrated in 1992 at Strasburg Observatory); the 26th Conference SEAC, in Graz (Austria), is scheduled for August–September 2018. The SIAC has organised six field schools and five workshops. The *VII Escuela* and *VI Jornadas Interamericanas de Astronomía Cultural*, titled *Agua y Cielo*, to be held in Samaipata (Bolivia), are scheduled for October 2018. The ISAAC, for its part, is in charge of organising the aforementioned *'Oxford' International Symposia* and the publication of *Archaeoastronomy. Journal of Astronomy in Culture* (https://escholarship.org/uc/jac); this journal, which is based in the University of California, is open access and is published twice a year, coinciding with the solstices.

[15]After being recognised as a scientific discipline, the next challenge is to have archaeoastronomy regularly incorporated into teaching plans, for instance in Spanish universities (Belmonte 2009: 65; Cerdeño et al. 2006: 25–26).

[16]Edited in three volumes by C.L.N. Ruggles (Heidelberg, 2014–2015), it presents an up-to-date theoretical and methodological perspective, as well as including thematic approaches centred on specific topics such as cosmologies, calendars, navigation, orientation and alignments and ancient perceptions of space and time; the work also includes ethnoastronomical studies which focus on current 'indigenous' groups and some wide-ranging geographical and chronological case studies.

World Heritage Sites[17] and their protection by international organisations such as UNESCO and ICOMOS.[18]

In this context, the 1990s also witnessed the emergence of veritable national schools of archaeoastronomy and cultural astronomy. I want to emphasise two of them, because of the prominence that their members have gained worldwide, and because of the relevant role that they play in this volume. Two milestones stress the interest for the discipline in Spain (Belmonte 2009: 55–67; Esteban 2003a: 309–322): M. Hoskin's study of the alignment of Iberian megaliths, from the 1980s onwards (Hoskin 2001),[19] and Jaschek's time in Salamanca (1993–1999), which was a boost for the discipline in Spain and led to the organisation of various seminars (e.g. *Astronomía y Ciencias Humanas*), among which the celebration of the 1996 SEAC annual meeting in Salamanca (1996) (Jaschek and Atrio Barandela 1997) may be highlighted. This favourable context also witnessed the formation of the first (and to date unparalleled) Spanish archaeoastronomy team[20] in the *Instituto de Astrofísica de Canarias* (IAC) and the University of La Laguna (Tenerife). This team was led by the astrophysicist J.A. Belmonte and included the physicists César Esteban and Antonio César González, among others.[21]

---

[17]See, for instance, the volumes published by the *International Council on Monuments and Sites* [ICOMOS] and the *International Astronomical Union* [IAU]: Ruggles (2017) and Ruggles and Cotte (2010).

[18]Archaeoastronomy will be considered a thematic area in the forthcoming *ICOMOS International Scientific Committee for Archaeological Heritage Management 2018 Annual Meeting*, to be celebrated in October 2018 in Montalbano Elicona (Sicily, Italy), under the title *Discover Sicily's Argimusco. A Holistic Approach to Heritage Management* (http://icahm.icomos.org/2018-icahm-annual-meeting-sicily/).

[19]Hoskins established important links with local teams, such as those led by M.ª L. Cerdeño and G. Rodríguez Caderot (University Complutense de Madrid), and M. García Quintela and F. Criado (University of Santiago de Compostela), and those which focused on Islamic astronomy (Belmonte 2009: 63) and the Iberian world (Esteban 2002: 81–100; and Espinosa Ruiz 2018: 265–278). See Cerdeño and Rodríguez (2009), *Arqueoastronomía* (*Complutum*, 20, 2), and especially the synthetic, conceptual, epistemological and methodological works by G. Rodríguez Caderot and M.ª L. Cerdeño Serrano, Stanislaw Iwaniszewski, Marco V. García Quintela, A. César González García and Juan Antonio Belmonte Avilés. For the international projects undertaken by these teams, see Lull (2006).

[20]http://www.iac.es/proyecto/arqueoastronomia/

[21]This team, which from the 1990s onwards undertook several projects in cooperation with other European and American colleagues, also organised the *VI 'Oxford' Symposium* (1999) in La Laguna and convened the organisation of the research group *Arqueoastronomía* within the framework of the IAC (Esteban and Belmonte 2000), whose main aim is to assess the role of astronomy in the cultural milieu of past civilisations, from prehistory to our days. The interests of the group go beyond the local perspective (Aparicio and Esteban 2005; Belmonte et al. 1995: 133–156), largely focusing on Mediterranean societies, from the Atlantic façade to the Middle East (Belmonte and Shaltout 2009), and especially the Iberian Peninsula. They have also carried out some work concerning Mesoamerican and Polynesian (Easter Island) societies. The prestigious work undertaken by this research group at the international level is of enormous importance; the analysis of such a wide variety of cultural horizons from an astronomical perspective involves the participation of experts with an in-depth knowledge of historical and archaeological sources as well as of the

In Italy, the interest in Sardinian dolmens, popularised by Hoskin in the 1980s (Hoskin and Zedda 1997: 1–16; Magli et al. 2011), progressively expanded to other regions (Puglia, Lazio, Veneto and Valle d'Aosta) (Aveni and Romano 1986: 23–31; Romano 1992). After a series of meetings convened by the *Accademia Nazionale dei Lincei*, a group of archaeologists, astronomers and practitioners of associated disciplines[22] created the *Società Italiana di Archeoastronomia* (SIA) in Milan in 2000.[23]

This volume is the result of the collaboration between Spanish and Italian scholars, which began in earnest during the *16th Conference of the Italian Society for Archaeoastronomy*, titled *Quis dubitet hominem coniungere caelo?*[24] As in previous meetings, the conference was a forum in which to continue exploring the relationship between the cosmos and human societies, from prehistory to our days. In addition, the organisers had—in my opinion—the felicitous idea of convening, in parallel with the main meeting, the *1st International Workshop on Archaeoastronomy in the Roman World*,[25] in response to an increasing interest in Classical, especially Roman, archaeoastronomy, over the previous decade.

In general, the current concept of Classical archaeology has transcended the limits of the Graeco-Roman cultural milieu. In this new social and chronological dimension, the field is also interested in the study of cultures that co-existed with the Classical civilisations, such as the Italian protohistoric societies and the Germanic peoples (Gutiérrez 1997: 51–52). Within this expanded discipline, Roman archaeology is now divided into multiple specialised fields (the diachronic study of the polity of Rome, the Italian Peninsula, the Eastern and Western provinces, etc.). It is, therefore, not unreasonable to demand the configuration of a specialised field, the aim of which would be to analyse the way Romans (and the societies that preceded and followed what we understand as Ancient Rome) related to the cosmos and

---

operation of social processes (see, for instance, the following synthetic works: Belmonte 1999; Belmonte and Hoskin 2002; Belmonte and Sanz de Lara 2001).

22 Including the archaeologist Gustavo Traversari and the astronomers Edoardo Proverbio, Giuliano Romano and Elio Antonello.

23 The association is based in the Osservatorio Astronomico di Brera and was created with the aim of promoting archaeoastronomy, ancient astronomy, cultural astronomy and historical astronomy. These aims emphasise the inherently interdisciplinary nature of the field (Antonello 2003: 507–513); see http://www.brera.inaf.it/archeo/index.htm.

24 The meeting was organised by the Department of Mathematics of the Politecnico di Milano (Italy) on 3–4 November 2016. The scientific committee was formed mainly from important members of the archaeastronomical communities in Italy and Spain and included E. Antonello, J.A. Belmonte, A.C. González-García, R. Hannah, M. Incerti, G. Magli, V.F. Polcaro and G. Rosada; see https://www.mate.polimi.it/sia2016/.

25 Astrophysicists linked to important research institutions (concerning such fields as physics, astrophysics and heritage studies) are currently consolidating the field of archaeoastronomy in the European continent. Their work is analysing the relationship between architecture, landscape and mathematic-astronomical knowledge in ancient societies (especially concerning European megaliths, the prehistoric, protohistoric and Roman Mediterranean, Egypt and the Near East).

astronomical phenomena through the analysis of astronomical, archaeological and historical data.[26]

In a paper published in 2006, Cerdeño et al. (2006) carried out a bibliometric analysis of the papers published by the journal *Archaeoastronomy* between 1979 and 2002. They concluded that 31.8% of the papers dealt with European megaliths, and only 6.2% focused on the Classical period (Cerdeño et al. 2006: 20). These results are hardly representative, but reflect an emphasis—from the beginning of the discipline—on megalithism, especially in Europe. In recent years, the situation has changed substantially; as previously noted, over the last two decades archaeoastronomical studies have become much more widespread, covering almost every past human society, including Roman civilisation.

This volume, edited by Giulio Magli, A. César González-García, Juan Belmonte Aviles and Elio Antonello, follows a threefold diachronic, geographical and thematic structure. It is divided into several sections, dealing with Etruria—the earliest Italian culture during the Iron Age—and the Roman Empire (first to fourth centuries AD); special subsections address the *Urbs*, other Roman cities, the Eastern provinces and the application of computer methods to archaeoastronomy, which have led to the emergence of a new discipline: virtual archaeoastronomy.

We know that the Roman *libri vegoienses* contained instructions for the interpretation of electrical phenomena (the *libri fulgurales* and especially the *libri rituales*).[27] Antonio P. Pernigotti, an archaeologist at the Università degli studi di Milano, reassesses matters of orientation and ritual among Etruscan temples, which have been previously examined by different authors (e.g. Aveni and Romano 1994: 545–563; Prayon 1991: 1285–1295). Pernigotti aims to examine whether the orientation of Etruscan temples was random or whether they followed any rules regarding order and proportion. After measuring the azimuth and, whenever possible, the horizon height of major Etruscan sacred structures (28 temples in 10 different locations, 9 in Etruria and 1 in Tuscia)—and finding errors in previous measurements—Pernigotti relates the data with the chronology of, and the deity worshipped in, each temple, in order to determine possible patterns in the orientation of the temples. Based on his results, Pernigotti argues that Etruscan temples were oriented according to the Sun, rather than to the celestial dwellings of the deities (known after Martianus and the famous Bronze Liver of Piacenza). With some exceptions, temple facades were not oriented towards the dawn, and their *cellae* were not illuminated by solar rays; instead, there seems to have been a function between orientation and specific deities (Uni, Vea, Hercules . . .).

[26]The inclusion of several Roman-centred case studies in the *Handbook of Archaeoastronomy and Ethnoastronomy* (e.g. González-García and Magli 2014: 1643–1650) demonstrates the consolidation of this discipline, which is also illustrated by this volume.

[27]According to Festus, these were "Etruscan books which prescribe rituals for the foundation of cities and the consecration of altars and temples, the blessing of walls and the norms to distribute city gates and organise tribes, curiae, and centuriae; to organise armies and all else that pertains to war and peace." (Festo, *Rituals*). See Bagnasco et al. (2013) for a recent account on how these texts might be related to the orientation of the sanctuary of Tarquinia.

G. Bagnasco Gianni (specialist in Etruscan epigraphy in Università degli studi di Milano) combines Etruscan and Roman rituals related to the foundation of cities (Briquel 2008: 27–47; Rykwert 1988) and some cosmological principles of the Etruscan religion to re-examine (Bagnasco Gianni and Facchetti 2015: 27–56; Bagnasco Gianni et al. 2016: 253–302) the Tumulus of the Crosses, in Cerveteri, in whose corridor an Orientalising inscription containing the names of various divinities inside a celestial quadrant and a *siglum* formed by a cross inside a circle was found. The pictogram is divided into 16 regions, one for each deity, as also reflected in the previously noted Liver from Piacenza. Based on the differences and similarities between the information conveyed by the Liver and other written sources (especially Pliny and Martianus), Bagnasco Gianni concludes that the north-western orientation of the wall associated with the access ramp, where the inscription was found, allows for the beginning of the sequence of divinities mentioned in the Liver (and Martianus) to be established in the north-eastern quadrant. As such, the division of the Liver which contains the expression *Tin Ciles*, to the east of the division which contains the expression *Ciles* alone, could signal the increase in sunlight (*Tin*) that follows the sunrise at the summer solstice.

Concerning the Early Imperial period, some attention has been paid over the last decade—by Magli, Belmonte and González-García, among others—to the astronomical orientation of cities and buildings, especially in Italy and the western provinces, in relation to rituals and government propaganda.[28] Deliberate sunlight effects, as a way to stress hierophanies, most prominently found in the northern sector of the *Campus Martius* and in the triangle formed by the *Ara Pacis*, the *Horologium* and Augustus' Mausoleum (Buchner 1976; Hasalberger, 2014; Rehak 2006), can be attested in many more constructions in both Rome[29] and other regions.[30] Three case studies are analysed in this volume—specifically, in the sections dedicated to the *Urbs*, virtual archaeology and archaeoastronomy. A team integrated by V.F. Polcaro, S. Sclavi, S. Gaudenzi, L. Labianca and M. Ranieri (from the Universities of Ferrara and Roma La Sapienza and the Soprintendenza Archeologica di Roma) reassess (Labianca et al. 2008) the study of the so-called Neo-Pythagorean basilica of Porta Maggiore, an underground complex dated to the first century AD and located in the city suburbs. The complex has been interpreted as being related to Neo-Pythagorean cults, or otherwise as the funerary mausoleum of the consul *T. Statilius Taurus*. The evening sunlight penetrated the complex through a skylight in the vault of the vestibule (this was especially intense around the summer solstice) and, less directly, through a window in the main nave, projecting a point of light upon an altar. The authors argue that, by placing a light-reflecting surface on the

[28]Bertarione and Magli (2015); Esteban (2003b); Espinosa-Espinosa and González-García (2017); Ferro and Magli (2012); García Quintela and González-García (2014: 157–177); González-García and García Quintela (2014); Magli (2008: 63–71); Magli et al. (2014); Rodríguez-Antón (2017).

[29]For instance, in the Neo-Pythagorean basilica in Porta Maggiore, the Mausoleo degli Equinozi in the Via Appia (De Franceschini 2012), the Pantheon (Hannah and Magli 2011) and Adrian's mausoleum (De Franceschini and Veneziano 2015).

[30]For an illustrative example from Spain, see the studies about the orientation of the sanctuary of Torreparadones (Córdoba) (Abril Hernández and Morena-López 2018: in press).

altar, this point of light would fall upon a painting interpreted (not without some doubt) as the 'Rape of Ganymede'. According to the authors, this hierophany would be at the centre of the rituals celebrated in the complex.

Based on an allegorical interpretation of a text by the Neo-platonic philosopher Porphyry, who—inspired by the Mithraic mysteries—assimilated the nymphs' cavern described by Homer (Od. 13.102–112) to the cosmos, R. Hannah (University of Otago, New Zealand) suggests that solar equinoxes were points of balance in which gods and (deified) emperors were enthroned and that northern and southern solstices opened 'passages' for the transit of souls. According to Hannah, these ideas found reflection in public architecture, in which domes were used to represent the cosmos. The Pantheon, rebuilt in the centre of the *Campus Martius* during Hadrian's reign, was originally an *augusteum* and a temple consecrated to the divine pantheon, but later it was also used to determine the agricultural cycle, and its interior operated as a giant sundial (*hemiciclum*); on the equinoxes, at noon, the Sun entered the building and fell above the entrance, and this also happened on anniversaries meaningful to Imperial propaganda (e.g. 21 April, the day of the foundation of Rome). This use of light can also be attested in Nero's *Domus Aurea*; several poetic and historical sources recount that the entrance to the Octagonal Room was hit by sunlight on the equinoxes, also at noon (Hannah et al. 2016). Some evidence suggests that these effects were similarly used in Domitian's palace, in the Palatine. Finally, Hannah tries to picture the celebration of these hierophanies, based on the analysis of a number of Eastern Byzantine churches; in these churches, geometry, light and cosmology were used to provide light effects for solemn parades and processions, as a symbol of the divine will. According to Hannah, these light-infused rituals could be inspired by, and provide evidence for, Imperial ceremonies, also known to have taken place in Villa Adriana (De Franceschini and Veneziano 2013).

Finally, the effect of sunlight on the Mausoleum of Santa Constanza, dated to the second half of the fourth century AD and located in the archaeological complex of Sant'Agnese fuori le Mura in Rome (Rasch and Arbeiter 2007), is examined for the first time, in this volume. Based on the Mausoleum's azimuth, calculated with satellite technology, Flavio Carnevale and Marzia Monaco, from the Università degli studi di Roma La Sapienza—who have also made interesting contributions to the study of the orientation of Greek theatres, Etruscan funerary *tholoi*, the *Portunnus* temple in the Forum Boarium of Rome and the Mithraea of Ostia Antica—have attested two phenomena: (1) after the construction of the skylight (late fourth century), the Sun illuminated the interior between 8 and 25 February, during the festival of the *Parentalia*; this heliophany was similar in character to that attested in the Pantheon every 21 April; and (2) the sunlight would hit directly the centre of the Mausoleum, where the porphyry sarcophagus of Costanza was originally located.

The foundation of Roman cities and buildings was rooted in mythology, and religious formulae were essential for the projection of the celestial order upon the landscape and the spaces ritually arranged by the magistrates (Rykwert 1988). Various research projects, especially in the western provinces, have analysed the orientation of cities (an issue already mentioned by various classical authors such as

Hyginus Gromaticus [*Constitutio*, I] and Frontinus (*De Agrimensura*, 27]), the methods used to calculate these orientations and their symbolic meaning. This method, along with other similar practices, became especially popular during Augustus' reign and has been thoroughly studied in Hispania and the western provinces by the members of the archaeoastronomy team of the IAC (Belmonte et al. 2016: 65–77; González-García 2015: 141–162; González-García, et al. 2014: 107–119). Several members of this team (A.C. González-García, A. Rodríguez-Antón, and J.A. Belmonte), in cooperation with D. Espinosa-Espinosa and M.V. García, analyse the landscape of western Augustan cities (measurements have been taken in 64 of these cities, in Hispania, Gaul, Germania, Italy and North Africa). These measurements are compared with the celestial landscape, and increasingly clear orientation patterns have begun to emerge. The evidence suggests a preference for orienting cities towards dawn on the winter solstice and, to a lesser extent, on the equinoxes and the summer solstice. The study also aims to define with more precision than has been possible hitherto the 'solar model' that identified Augustus with the Sun and Apollo, a relationship that the *princeps* used regularly for propaganda purposes. Also in relation to this, several of these authors (A. Rodríguez-Antón, A.C. González-García and J.A. Belmonte), this time in cooperation with M. Orfila (archaeologist at the University of Granada), address the use of the geometric measurement technique known as *varatio* (Orfila et al. 2014). Based on measurements taken on 81 Iberian cities, it is concluded that this technique may have been used to calculate azimuths. However, the authors admit that, although it may be argued that *varatio* was used in order to organise urban and rural landscapes, which would indicate a direct link between technique and symbol when celestial phenomena were not available for direct observation, a larger sample of case studies is necessary.

The eastern provinces, which were thoroughly Hellenised and which maintained strong links with the Near East, have also been analysed from an archaeoastronomical perspective. These studies, which have been particularly intense in the Mediterranean Levant and Egypt, generally focus on the orientation of architectural features and the occupation of the landscape. Again, the team from the IAC, led by Belmonte, González-García and Rodríguez-Antón, analyse the so-called Khirbet et-Tannur Zodiac (Hurawa), which they suggest should be relabelled as an 'almanac' or 'parapegma'. This feature was found on the main altar of the sanctuary of Djebel Tannur (Jordan) (*Arabia Adquisita*) and depicts an impressive astral cycle dated to the Roman period. Despite the persistence of Nabatean traditions, Roman domination led to the adoption of the Julian calendar in the region, although the different months were still named after the lunisolar Nabatean calendar. This is the origin of the so-called *Era Provincia Arabia*. In this volume, the feature is interpreted as a calendric guide to the rituals celebrated in the sanctuary, and the measurements indicate that it was oriented towards dawn three days prior to the spring equinox, on 22 March, that is, day 1 of Nisan (New Year's Day in the Arabian Calendar). This could mean that the complex was regarded as a national sanctuary and the destination of the pilgrimage of Djebel Tannur. A small temple in Petra, erected between AD 106 and 114, and dedicated to the imperial cult, presents the same orientation (Belmonte and González-García 2017), suggesting the capacity of the Nabateans to adapt to the Roman domination.

In the following chapter, G. Magli undertakes the study of the orientation (which has not been measured to date) of the temple *podia* of the megalithic sanctuary of Jupiter in Heliopolis (Baalbek, Lebanon) (Kropp 2009; 2010; Segal 2013). The Sun aligns with the temple on 1 May and 12 August, dates with no special implications in the solar cycle, which could challenge the solar associations of the temple. However, Magli's results indicate that the temple was aligned with the Pleiades (the Seven Sisters) during the reign of Herod the Great, which could confirm the temple's association with Jupiter and the agricultural cycle. Magli also suggests that both *podia* may have been built at the same time, during the reign of Herod the Great.

C. Rossi and G. Magli undertake an analysis of Late Roman fortified settlements in Egypt's Western Desert (Rossi and Ikram 2018), which illustrate how Romans interacted with the landscape, following well-known precepts by such authors as Pliny and Vitruvius. These settlements seem to have been oriented towards the dominating winds, from the north-west, the azimuth of which was calculated by measuring the axial axles of the surrounding sand dunes. It is unclear whether this 'weathervane orientation' responds to pre-Roman traditions or whether it answers to astronomic concerns and the desire to adapt as much as possible to local environmental conditions (wind, topography, etc.).

Finally, two chapters analyse the use of computer applications and virtual archaeology, a useful combination for the analysis and dissemination of the historical and archaeological features of ancient buildings. The first of these chapters addresses the virtual reconstruction and archaeoastronomical analysis of the Mausoleum of Theoderic, in Ravenna, built during Theodric's reign and heavily transformed from the eighth century onwards. After examining the original elements, M. Incerti, G. Lavoratti and S. Iurilli (architects at the Università degli studi di Ferrara) explain the development of a 3D model of the exterior and interior of the building and use its astronomical orientation (measured in 1995 by G. Romano) to analyse the effects of sunlight at different times of the year. Specifically, light entered through some of the windows on the solstices, illuminating important elements in the interior, including the red porphyry sarcophagus. As noted in the introduction to R. Hannah's chapter, above, this kind of light effect was a common way to highlight special dates or times of the day. In addition, the geometrical study of the Mausoleum revealed that, rather than the mathematical calculations conveyed by *De Geometria*, attributed to Boethius, the architect of the Mausoleum used the geometrical tools contained in Euclid's *Elements.* The results also suggested that the builders of this monument occasionally used the Roman foot, although the standard measurement unit in the building was the Byzantine foot.

Finally, G. Zotti, B. Frischer, F. Schaukowitsch, M. Wimmer and W. Neubauer, who work on generating 3D models of archaeological buildings and features (Ludwig Boltzmann Institute for Archaeological Prospection and Virtual Archaeology de Vienna, Indiana University and the Institute of Computer Graphics and Algorithms of Vienna), are currently giving the final touches to the software *Stellarium*, which will present these buildings and features in relation to celestial bodies. The result of this project is the so-called *Skyscape Planetarium*, which presents the relationship between the Earth's landscape and the sky in an accessible way. The tool can simulate the astronomic relationships of buildings with total precision, showing orientation patterns and sunlight and moonlight effects (Zotti 2015).

These tools, which were publicly displayed with great success in the exhibit *Stonehenge. A Hidden Landscape*, celebrated in 2016–2017 in the MAMUZ Museum (Mistelbach, Austria), have also been applied in two 'Roman' projects: (1) the study of the astronomical orientation of the *Antinoieon*, in Villa Adriana (Frischer et al. 2016: 55–79), which has confirmed Mari's initial theses (Mari and Sgalambro 2007: 83–104) that the so-called Temple 1 is oriented towards dawn, on the summer solstice (festival of *Fors Fortuna*), and towards the constellation of Antinoo, the heliacal configuration of which occurs around the birthday of Hadrian's unfortunate lover, and (2) the analysis of the relationship between the *Ara Pacis* and the *Horologium* (Frischer et al. 2017: 18–119; Frischer 2017–18, 3–100), which has identified the mistakes upon which Buchner's (1976: 19–65) famous hypothesis was built, and has determined that the shadow of the obelisk did not run along the equinoctial line across the central area of the altar on Augustus' birthday; the emergence of the Sun over the top of the obelisk, on the other hand, has been confirmed, which reinforces the idea that the obelisk was dedicated to the Sun, already suggested by the epigraphic evidence.

There is little doubt that these works will notably contribute to the consolidation and dissemination of Roman archaeoastronomy, while highlighting the central role played by astronomical observation and celestial phenomena for the Romans, a key factor for the symbolic and mythical aspects of their society.

Universidad de Murcia, Murcia, Spain José Miguel Noguera Celdrán
15 June 2018

References

Abril Hernández, J. M., & Morena-López, J. A. (2018). Archaeoastronomical study on the Iberian-Roman Sanctuary of Torreparedones (Baena, Spain). *Journal of Skyscape Archaeology, 4*(1), in press.

Antonello, E. (2003). Cultural astronomy and archaeoastronomy: An Italian experience. In I. Šprajc, & P. Pehani (Eds.), *Ancient cosmologies and modern prophets, Proceedings of the 20th conference of the European Society for Astronomy in Culture* (*Anthropological Notebooks*, year XIX, supplement) (pp. 507–513). Ljubjana: Slovene Anthropological Society.

Antonello, E. (2012). L'astronomia culturale e l'archeoastronomia dai convegni lincei a oggi. In *Convegno lo Sguardo sugli Astri. Scienza, Cultura e Arte*, 2 Aprile 2012 (pp. 163–172). Roma: Accademia Nazionale dei Lincei (http://www.

brera.inaf.it/utenti/antonello/AntonelloTestoLincei.pdf, consultado el 4 de junio de 2018).

Aparicio, A., & Esteban, C. (2005). *Las Pirámides de Güimar. Mito y Realidad*. La Laguna: Centro de Cultura Popular Canaria.

Aveni, A. F. (1991). *Observadores del Cielo del Antiguo México*. México: Fondo de Cultura Económica.

Aveni, A. F., & Romano, G. (1986). Archaeoastronomical research in Veneto-Friuli, Italy. *Archaeoastronomy, Supplement of the Journal for the History of Astronomy, 10*, S23–S31.

Aveni, A., & Romano, G. (1994). Orientation and Etruscan ritual. *Antiquity, 68*, 545–563.

Bagnasco, G. G., Bortolotto, S., Favino, P., Garzulino, A., Marzullo, M., Riva, E., Simonelli, R., Valtolina, S., & Zerboni, A. (2013). Past&Present at Tarchna&Tarquinia: A flexible approach to make visible the invisible. In M. Boriani, R. Gabaglio, & D. Gulotta (Eds.), *Built Heritage 2013 – Monitoring, conservation and management* (pp. 163–175). Proceedings of the international conference (Milano 2013). Milano: Politecnico di Milano – Centro per la Conservazione e Valorizzazione dei Beni Culturali.

Bagnasco Gianni, G., & Facchetti, G. M. (2015). Considerazioni sulla rilettura di CIES5992. *Alessandria*, 8, 27–56.

Bagnasco Gianni, G., Gobbi, A., & Scoccimarro, N. (2016). Segni elocuenti in necropoli e abitato. In M. L. Haack (Ed.), *L'Écriture et l'Espace de la Mort. Epigraphie et Nécropoles à l'Époque Prérromaine* (pp. 253–302). Roma: École Française de Rome.

Bauer, B. S., & Dearborn, D. S. (1998). *Astronomía e Imperio en los Andes*. Cusco: Centro de Estudios Regionales "Bartolomé de las Casas".

Belmonte, J. A. (co-ord.) (1994; 2000^2). *Arqueoastronomía Hispánica. Prácticas Astronómicas en la Prehistoria de la Península Ibérica y los Archipiélagos Balear y Canario*. Madrid: Equipo Sirius.

Belmonte, J. A. (1999). *Las Leyes del Cielo*. Madrid: Temas de Hoy.

Belmonte, J. A. (2009). La arqueoastronomía en Europa. La singularidad del caso Español. *Complutum, 20*(2), 55–67.

Belmonte, J. A. (2016). Orientatio ad Sidera (Oas). Highlights of a decade of archaeoastronomical research in the Mediterranean region and beyond. Mediterranean Archaeology and Archaeometry, *16*(4), 93–101.

Belmonte, J. A., & González-García, A. C. (2017). Petra re-visited: New clues for the Nabataean cultic calendar. *Culture and Cosmos*, in press.

Belmonte, J. A., & Hoskin, M. (2002). *Reflejo del Cosmos. Atlas de Arqueoastronomía del Mediterráneo Antiguo*. Madrid: Equipo Sirius.

Belmonte, J. A., & Sanz de Lara, M. (2001). *El Cielo de los Magos*. Santa Cruz de Tenerife: La Marea.

Belmonte, J. A., & Shaltout, M. (Eds.), (2009). *In search of cosmic order. Selected essays on Egyptian archaeoastronomy*. Cairo: Supreme Council of Antiquities Press.

Belmonte, J. A., Esteban, C., Aparicio, A., Tejera, A., & González, O. (1995). Canarian astronomy before the conquest: The pre-Hispanic calendar. *Revista de la Sociedad Canaria para la ciencia, VI*, 133–156.

Belmonte, J. A., Noguera Celdrán, J. M., González-García, A. C., & Rodríguez-Antón, A. (2016). Astronomy and landscape in Carthago Nova. In F. Silva, K. Malville, T. Lomsdalen, & F. Ventura (Eds.), *The materiality of the sky. The proceedings of the twenty-second annual conference of the Societé Européenne pour l'Astronomie dans la Culture (SEAC) held in Malta.* 22–26 September 2014 (pp. 65–77). Wales: Sophia Center Press.

Bertarione, S. V. & Magli, G. (2015). Augustus' power from the stars and the foundation of *Augusta Praetoria Salassorum. Cambridge Archaeological Journal, 25*(1), 1–15.

Bianchi Bandinelli, R. (1992[2]). *Introducción a la Arqueología Clásica como Historia del Arte Antiguo*. Madrid: Ediciones Akal (edic. española).

Briquel, D. (2008). L'espace consacré chez les Étrusques: reflexions sur le rituel Étrusco-Romain de fondation des cites. In X. Dupré Raventós, S. Ribichini, & S. Verger (Eds.), *Saturnia Tellus. Definizioni dello Spazio Consacrato in Ambiente Etruco, Italico, Fenicio-Punico, Iberico e Céltico* (pp. 27–47). Rome: Consiglio Nazionale delle Ricerche.

Buchner, E. (1976). Solarium Augusti und Ara Pacis. *Römische Mitteilungen, 83*, 19–65.

Cerdeño Serrano, M. L., & Rodríguez-Calderot, G. (2009). La arqueoastronomía ¿una arqueometría más? In M. E. Sáiz Carrasco, R. López Romero, M. A. Cano Díaz-Tendero, & Calvo García, J. C. (Eds.), *VIII Congreso Ibérico de Arqueometría* (pp. 279–286). Teruel: Seminario de Arqueología y Etnología Turolense.

Cerdeño Serrano, M. L., Rodríguez-Caderot, G., Moya, P. R., Ibarra, A., & Herrero, S. (2006). Los estudios de arqueoastronomía en España: estado de la cuestión. *Trabajos de Prehistoria, 63*(2), 13–34.

Chamberlain, V. D., Carlson, J. B., & Young, M. J. (Eds.), (2005). Songs from the sky. Indigenous astronomical and cosmological traditions of the World. In *Selected proceedings of the "first international conference on ethnoastronomy. Indigenous astronomical and cosmological traditions of the World"* held at the Smithsonian Institution, Washington, DC, 5–9 September 1983. Bognor Regis: Ocarina Books.

Cunliffe, B., & Renfrew, C. (co-ords.) (1997). *Science and Stonehenge.* Oxford: Oxford University Press.

De Franceschini, M. (2012). Orientamenti equinoziali nel Mediterraneo: il tempio preistorico di Mnajdra a Malta e il Mausoleo degli Equinozi a Roma. In *XIV Seminario di Archeoastronomia* (Associazione Ligure per lo Sviluppo degli Studi Archeoastronomici) (pp. 18–29). Genova: Associazione Ligure per lo Sviluppo degli Studi Archeoastronomici.

De Franceschini, M., & Veneziano, G. (2013). Architecture and archaeoastronomy in Hadrian's villa near Tivoli, Rome. *Nexus, 15*(3), 457–485.

De Franceschini, M., & Veneziano, G. (2015). Archeoastronomia nell Mausoleo di Adriano, Castel Sant'Angelo (Roma). In *XVII Seminario di Archeoastronomia* (Associazione Ligure per lo Sviluppo degli Studi Archeoastronomici) (pp. 5–17). Genova: Associazione Ligure per lo Sviluppo degli Studi Archeoastronomici.

Espinosa-Espinosa, D., & González-García, A. C. (2017). A.D. VIIII Kalendas Octobres, Dies Natalis Augusti. Some considerations on the astronomical orientation of Roman Cologne and the Imperial cult. *Numen, 64*(5–6), 545–567.

Esteban, C. (2002). Elementos astronómicos en el mundo religioso y funerario Ibérico. *Trabajos de Prehistoria, 59*, 2, 81–100.

Esteban, C. (2003a). La arqueoastronomía en España. *Anuario del Observatorio Astronómico de Madrid,* 309–322.

Esteban, C. (2003b). Temples and astronomy in Carthage. In *Calendars, symbols, and orientations. Legacies of Astronomy in Culture* (pp. 135–142). Uppsala: Uppsala Astronomy Observatory Reports, 59.

Esteban, C. (2009). La astronomía cultural, ¿es interdisciplinar? Reflexiones de un astrofísico. In M. L. Cerdeño, & G. Rodríguez Caderot (Eds.), A*rqueoastronomía* (*Complutum* 20, 2), 69–77.

Esteban, C., & Belmonte, J. A. (2000). *Astronomy and cultural diversity. VI Oxford Symposium and VIIth Conference SEAC, La Laguna, 1999*. Tenerife: Cabildo de Tenerife (una selección de los trabajos aquí presentados en: *Archaeoastronomy. The Journal of Astronomy in Culture, 15, 2000*).

Esteban, C., & Espinosa Ruiz, A. (2018). El equinoccio en el ritual Ibérico. El santuario de La Malladeta (La Vila Joiosa, Alicante). *Archivo Español de Arqueología, 91*, 265–278.

Ferro, L., & Magli, G. (2012). The astronomical orientation of the urban plan of Alexandria. *Oxford Journal of Archaeology, 31*, 381–389.

Frischer, B. (2017–2018). Edmund Buchner's Solarium Augusti: New observations and simpirical studies with technical appendices by Paolo Albèri Auber, David Dearborn, John Fillwalk, Mika Kajava, and Stefano Floris). *Rendiconti della Pontificia Accademia Romana di Archeologia, 89*, 3–100.

Frischer, B., Zotti, G., Mari, Z., & Capriotti Vittozzi, G. (2016). Archaeoastronomical experiments supported by virtual simulation environments: Celestial alignments in the Antinoeion at Hadrian's villa (Tivoli, Italy). In *Digital Applications in Archaeology and Cultural Heritage*, *3*, 55–79.

Frischer, B., Pollini, J., et al. (2017). New light on the relationship between the Montecitorio Obelisk and the Ara Pacis of Augustus. *Studies in Digital Heritage, 1*, 18–119.

Galindo, J. (1994). *Arqueoastronomía en la América Antigua*. Madrid: Grupo Sirius.

García Quintela, M. V., & González-García, A. C. (2014). Le 1er Août à *Lugdunum* sous l'Empire Romain: bilans et nouvelles perspectives. *Revue Archáeologique de l'Est, 63*, 157–177.

González-García, A. C., & García Quintela, M. V. (2014). The 1st of August at Lugdunum: Astronomy and imperial cult in Galia. *Mediterranean Archaeology and Archaeometry, 14*, 3, 83–91.

González-García, A. C., & Magli, G. (2014). Roman city planning and spatial organization. In C. L. N. Ruggles (Ed.), *Handbook of archaeoastronomy and ethnoastronomy* (pp. 1643–1650). Heidelberg: Springer.
González-García, A. C., Rodríguez-Antón, A., & Belmonte, J. A. (2014). The orientation of Roman towns in Hispania: Preliminary results. *Mediterranean Archaeology and Archaeometry, 14*(3), 107–119.
González-García, A. C., Noguera, J. M., Belmonte, J. A., Rodríguez, A., Ruiz, E., Madrid, M. J., Zamora, E., & Bonnet, J. (2015). Orientatio ad Sidera: astronomía y paisaje urbano en Qart Hadašt/Carthago Nova. *Zephyrus, 75*, 141–162.
Gutiérrez Lloret, S. (1997). *Arqueología. Introducción a la historia material de las sociedades del pasado*. Alicante: Publicaciones Universidad de Alicante.
Hannah, R., & Magli, G. (2011). The role of the Sun in the Pantheon's design and meaning. *Numen, 58*, 486–513.
Hannah, R., Magli, G. & Palmieri, A. (2016). Nero's "solar" kingship and the architecture of the Domus Aurea. *Numen, 63*(5–6), 511–524.
Haselberger, L. (2014). *The horologium of Augustus. Debate and context*, Supplementary Series, 99. Portsmouth, RI: Journal of Roman Archaeology.
Hawkins, G. S. (1973). *Beyond Stonehenge*. New York: Harper & Row.
Hoskin, M. (2001). *Tombs, temples and their orientations: A new perspective on mediterranean prehistory*. Bognor Regis: Ocarina Books.
Hoskin, M., & Zedda, M. (1997). Orientation of Sardinian dolmens. *Archaeoastronomy, 22*, 1–16.
Iwaniszewski, S. (1994). De la arqueoastrología a la astronomía cultural. *Trabajos de Prehistoria, 51*(2), 5–20.
Iwaniszewski, S. (1997). Archaeoastronomy and cultural astronomy: Methodological issues. In *Archeologia e Astronomia, Esperience e Propettive Future, ACL, 121* (pp. 17–26), Roma.
Iwaniszewski, S. (2003). The erratic ways of studying astronomy in culture. In M. Blomberg, P. E. Blomberg, & G. Henriksson (Eds.), *Calendars, symbols and orientations: Legacies of astronomy in culture*. In *Proceedings of the 9th Annual Meeting of SEAC*. Stockholm, 27–30 August 2001 (pp. 7–10). Uppsala: Mary Blomberg.
Iwaniszewski, S. (2009). Por una astronomía cultural renovada. In M. L. Cerdeño, & G. Rodríguez Caderot (Eds.), *Arqueoastronomía. Complutum, 20*(2), 23–37.
Jaschek, C., & Atrio Barandela, F. (Eds.), (1997). *Astronomía en la Cultura. Actas del IV Congreso de la SEAC*. Salamanca: Universidad de Salamanca.
Kropp, A. J. M. (2009). The cults of Ituraean Heliopolis (Baalbek). *Journal of Roman Archaeology, 22*, 365–380.
Kropp, A. J. M. (2010). Jupiter, Venus and Mercury of Heliopolis (Baalbek). The images of the "triad" and its alleged syncretisms. *Syria, 87*, 229–264.
Krupp, E. C. (1989). *En Busca de las Antiguas Astronomías*. Barcelona: Pirámide.
Krupp, E. C. (1991). *Beyond the Blue Horizon*. Oxford: Oxford University Press.
Labianca L., Sciortino I., Gaudenzi S, Patané A., Polcaro V. F., & Ranieri M. (2008). An archaeoastronomical study of the neo-Pythagorean basilica at Porta Maggiore in Rome. In *Proceedings of the SEAC 2007/Oxford VIII Conference Astronomy and Cosmology in Folk Tradition and Cultural Heritage. Archaelogia Baltica, 10*, 195–199.

Lehoux, D. (2007). *Astronomy, weather, and calendars in the ancient world: Parapegmata and related texts in classical and Near Eastern Societies*. Cambridge: Cambridge University Press.

Lull, J. (Ed.), (2006). *Trabajos de Arqueoastronomía. Ejemplos de África, América, Europa y Oceanía*. Gandía: Agrupación astronómica de La Safor.

Magli, G. (2008). On the orientation of Roman towns in Italy. *Oxford Journal of Archaeology, 27*(1), 63–71.

Magli, G., Realini, E., Sampietro, D., & Zedda, M. P. (2011). The megalithic complex of Monte Baranta in Sardinia: A pilgrimage center of the early Bronze Age? *Complutum, 22*(1), 107–116.

Magli, G., Realini, E., Reguzzoni, M., & Sampietro, D. (2014). Uncovering a masterpiece of Roman engineering: The project of Via Appia between Colle Pardo and Terracina. *Journal of Cultural Heritage, 15*(6), 665–669.

Mari, Z., & Sgalambro, S. (2007). The Antinoeion of Hadrian's villa: Interpretation and architectural reconstruction. *American Journal of Archaeology, 111*, 83–104.

Morellato, J. (2011). *Archeologia e Astronomía. Trecento anni d'Incontro*. Roma: Scienze e Lettere.

Orfila, M., Chávez-Álvarez, E., & Sánchez, E. H. (2014). *La Orientación de las Estructuras Ortogonales de Nueva Planta en Época Romana. De la Varatio y sus Variaciones*. Granada: Universidad de Granada and Universidad de La Laguna).

Prayon, F. (1991). *Deorum sedes*. Sull'orientamento dei temple Etrusco-Italici. *Archeologia Classica, 43*, 1285–1295.

Rasch, J. J, & Arbeiter, A. (2007). *Das Mausoleum der Constantina in Rom. Mit Beiträgen von Friedrich Wilhelm Deichmann und Jens Rohmann*. Mainz am Rhein: von Zabern.

Rehak, P. (2006). *Imperium and cosmos. Augustus and the Northern Campus Martius*. Madison: University of Wisconsin Press.

Rodríguez-Antón, A. (2017). *Cosmovisión y Urbanismo en la Roma Antigua: Orientación de Ciudades y Campamentos Romanos*. PhD thesis, Universidad de La Laguna.

Romano, G. (1992). *Archeoastronomia Italiana*. Padova: CLEUP.

Romano, G. (1995). *Orientamenti ad Sidera. Astronomia, Riti e Calendari per la Fondazione di Templi e Città. Un Esempio a Ravenna*. Ravenna: Edizioni Essegi.

Rossi C., & Ikram S. (2018). *North Kharga Oasis Survey. Explorations in Egypt's Western Desert*. British Museum Publications on Egypt and Sudan 5. Leuven: Peeters.

Ruggles, C. L. N. (1999). *Astronomy in prehistoric Britain and Ireland*. Princeton: Princeton University Press.

Ruggles, C. L. N. (2001). L'uso dell'archeoastronomía nell'esplorazione delle cosmología antica: probleme di teoría e método. *L'Uomo Antico e il Cosmos, ACL*, 171, Roma, 7–33.

Ruggles, C. L. N. (2014–2015). *Handbook of archaeoastronomy and ethnoastronomy*. New York: Springer.

Ruggles, C. (Ed.), (2017). *Heritage sites of astronomy and archaeoastronomy in the context of the World Heritage Convention. Thematic Study no. 2*. Paris: ICOMOS.

Ruggles, C., & Cotte, M. (Eds.). (2010). *Heritage sites of astronomy and archaeoastronomy in the context of the World Heritage Convention: A Thematic Study*. Paris: ICOMOS.

Rykwert, J. (1988). *The idea of a town: The anthropology of urban form in Rome, Italy and the ancient World.* Princeton: Princeton University Press.

Segal, A. (2013). *Temples and sanctuaries in the Roman East: Religious Architecture in Syria, Iudaea/Palaestina and Provincia Arabia*. Oxford: Oxbow Books.

Silva, F., & Campion, N. (Eds.), (2015). *Skyscapes: The role and importance of the sky in archaeology*. Oxford: Oxbow Books.

Šprajc, I. (2001). *Orientaciones Astronómicas en la Arquitectura Prehispánica de México*. México: INAH.

Thom, A. (1954). The solar observatories of megalithic man. *Journal of the British Astronomical Association, 64*, 396–404.

Thom, A. (1967). *Megalithic sites in Britain*. Oxford: Oxford University Press.

Thom, A., & Thom, A. S. (1984). Two major megalithic observatories in Scotland. *Archaeoastronomy, 7*, 129–148.

Zotti, G. (2015). Visualization tools and techniques. In C. L. N. Ruggles (Ed.), *Handbook for archaeoastronomy and ethnoastronomy* (pp. 445–457). New York: Springer.

Preface

In the last decade, there was increasing interest in the archaeoastronomy of the Roman epoch and many researches were carried out on this topic. Several studies have been devoted to the possible astronomical orientation of buildings—including the light effects with symbolic meaning—to the astronomical symbolism of the artefacts, and to the possible astronomical orientation of towns. Therefore, the time was more than ripe for a meeting dedicated specifically to 'Roman' archaeoastronomy. In 2016, the opportunity was offered by the Politecnico of Milan, Department of Mathematics, where the International Meeting on the Archaeoastronomy in the Roman World took place from 3 to 4 November. The meeting was followed by the Annual (XVIth) Conference of the Italian Society for Archaeoastronomy (SIA).

This volume includes a selection of the papers presented in that event organised into parts and chapters. Part I is devoted to the Etruscan Civilisation, from which Romans took several ideas. This includes two chapters that centre on an analysis of temples and on how the cosmology of the Etruscans was related to their funerary customs. In Part II, the orientation and solar light effects at given dates are considered for the Imperial buildings in Rome. One paper deals with the motivations of the symbolic use of equinoxes and solstices during the Imperial period, and two chapters illustrate the possible light effects in two monuments: the Basilica of Porta Maggiore and the Mausoleum of Santa Costanza. Part III is dedicated to the orientation of Roman towns, with a paper on a statistical analysis of a large set of sites, from central Europe to northern Africa, while another paper describes the possible use of a practical geometrical tool for planning the orientation of an urban grid. Two chapters in Part IV illustrate the astronomy in the provinces of the Empire under the influence of Roman rule. The first chapter proposes a new interpretation of the Tannur Zodiac (Nabataea) as a 'parapegma', and the second one discusses the chronology of the complex realisation of the Temple of Jupiter at Baalbek. The case of the Kharga oasis in Egypt is discussed in the third chapter, where the importance of the wind direction for the settlement is pointed out, while, in Part V, the first chapter is dedicated to the architectural and geometrical analysis of the Mausoleum of Theodoric. Finally, the second chapter in Part V describes the capabilities of the open source system Virtual Archaeoastronomy within the Stellarium software, intended for research and outreach, and shows some examples of its application to Roman monuments.

Etruscans, people who '. . . excelled everyone in religious observance . . .' (Livy 5.1.6), had a strong influence on political ideology and related rituals in the Roman world. The *Libri Rituales* of the *Etrusca Disciplina* also included the *Libri Fatales,* which probably contained the Etruscan founding rituals of cities and temples adopted by the Romans. The Etruscans gave great importance to the exact location of the cardinal points. According to Hyginus Gromaticus (first to second century CE), those points were the paradigm also for Romans, as regards at least (theoretical) land division. In the practical realisation, however, the Romans quite often adopted other criteria. For example, they usually took into account the physical characteristics of the places (e.g. rivers), or they used a simplified procedure to determine the East direction (and Hyginus showed why it could be erroneous). Vitruvius (first century BC), on the other hand, suggested a practical criterion for the town orientation based on the wind direction, since he was concerned mainly with the healthiness of the inhabitants, while he maintained the cardinal orientation for temples, when possible.

Things probably changed in part when Augustus introduced the (solar) cult of the Emperor. One may note in passing that he began this process with the divinisation of Julius Caesar in 44 BC by exploiting also an astronomical phenomenon, a comet that happened to appear during the period of the obsequy. It may be possible that astronomical criteria based on the sunrise at *specific* dates were then adopted for towns, temples and buildings. Many towns may be considered in this respect, since, as declared in the *Res Gestae*, Augustus settled colonies in Africa, Sicily, Macedonia, Spain, Achaea, Asia, Syria, Gallia Narbonensis and Pisidia, while Italy had 28 colonies founded under his auspices. Unfortunately, Augustus did not include the list, and historians tried several times to identify them (see e.g. Mommsen). As shown by inscriptions (e.g. OGIS 458), Oriental populations of the Roman Empire were keen to worship the Emperor (*Sebastos*). The positive attitude towards his divinity increased during the first centuries CE; for example, there are Roman coins with the representation of the Emperor as *sol invictus*. Several researchers have pointed out the light effects corresponding to solstice and equinox dates in Roman buildings of this epoch and connected in some way with the Emperor. Light effects based on specific astronomical orientations probably also were adopted later, but of course with a different meaning, for the Christian buildings.

A subtitle of the 'Joint 16th Conference of SIA and 1st International Workshop on Archaeoastronomy in the Roman World' was a quotation from Manilius' poem *Astronomica* (II.105)

> Quis dubitet [post haec] hominem coniungere caelo?,

that is, who can doubt that a link exists between heaven and man? Manilius was contemporaneous with Augustus and Ovid. In his poem, he described celestial phenomena, the zodiac and the related astrology. He and presumably many other people thought about cosmic harmony and the immanent divinity of nature. He wrote that into the soul of man God descends and seeks Himself and that the love of heaven makes us heavenly. Given that strong belief, it is therefore reasonable, not to say obvious, that present-day researchers would attempt to detect the expressions of such an astronomical link in ancient Roman artefacts and architecture, putting their results in archaeological and historical context. One can expect therefore further progress in this field.

Sadly, during the editing of these proceedings, we got the dismaying news that our colleague and friend Vito Francesco Polcaro passed away. Francesco was a polymath. He got three degrees, in mechanical engineering, aerospace engineering and mathematics; his scientific researches were mostly in high-energy astrophysics and technology, and in the astrophysics of the highest mass stars, but he also had deep interest in cultural astronomy, archaeoastronomy and archaeology. He collaborated with many professional and amateur archaeologists in the study of the astronomical content of ancient sites and artefacts, particularly in Rome and in Southern Italy. He regularly attended SIA and SEAC meetings, and he contributed actively to the organisation of several of our conferences. He was an enthusiastic man who believed passionately in science-led regulation and in the importance of the social aspects of science. Many people in primary and secondary schools and in cultural associations enjoyed his brilliant outreach talks. This volume on archaeostronomy in the Roman World is dedicated to the memory of Francesco.

As a final acknowledgement, we warmly thank the Politecnico of Milan for their hospitality and help with the organisation of the conference.

Milan, Italy Elio Antonello

Contents

Part V Roman Virtual Archaeology and Archaeoastronomy

Part I
Etruscan Temples and Cosmology

Chapter 1
A Contribution to the Study of the Orientation of Etruscan Temples

Antonio Paolo Pernigotti

Abstract The aim of the present contribution is to analyse the orientation of Etruscan temples in order to understand whether they followed a certain ratio or whether they were randomlyoriented. After a critical analysis of the data from azimuth measurements, an attempt is made to understand whether this ratio could depend on factors that are of ritualistic character, possibly linked to the sky and precepts of the *Etrusca Disciplina* which are known to us. In undertaking this study, the author has decided first to give a brief summary of previous studies on the present subject. In the final section of this paper, the author presents the data (collected in a series of field visits during the year 2013) pertaining to the azimuths of Etruscan temples, and furthermore makes some comments on the data analysis and its significance.

This contribution is part of the research project I am carrying out within the framework of the 'Tarquinia Project' of the Università degli Studi di Milano, designed since its beginning to carry out global archaeological, cultural and historical research. The project started in 1982 when Maria Bonghi Jovino began the excavations in one of the most ancient Etruscan sanctuaries, currently known as the 'monumental complex' and in the Ara della Regina sanctuary (*Tarchna* I 1997; *Tarchna* II 1999; *Tarchna* III 2001; *Tarchna* IV 2012). Over the past 10 years Giovanna Bagnasco Gianni, the present Director of excavation and research, has involved groups from the Università degli Studi di Milano and the Politecnico di Milano in order to gather more information about one of the most important aspects of Etruscan civilization, which is the design of space in accordance with the norms of Etruscan religion. In order to try and interpret such an invisible aspect by means of archaeological remains, cooperation with Giulio Magli started in 2012 and led to a first contribution on the orientation of the altar of the Ara della Regina sanctuary (Bagnasco Gianni et al. 2013). In the same year I started my research on the material aspects of the Etruscan sanctuaries from the point of view of the archaeoastronomical information given by the ancient Greek and Roman literary sources, which I am going to present in this paper.

A. P. Pernigotti (✉)
Università degli Studi di Milano, Milan, Italy
e-mail: antonio.pernigotti@unimi.it

G. Magli et al. (eds.), *Archaeoastronomy in the Roman World*, Historical & Cultural Astronomy, https://doi.org/10.1007/978-3-319-97007-3_1

Previous Studies on the Orientation of Etruscan Temples

The starting point of this paper is a critical analysis of what has been written before on the topic of the orientation of Etruscan temples.

In 1957, Ragna Enking first raised this question in a short essay where she came to the conclusion that the orientation of Etruscan temples was dependent upon the celestial sections inhabited by deities, with the frontal space of the buildings pointed towards the celestial dwellings of the deities worshipped there (Enking 1957).

In 1991, Friedhelm Prayon was the first to revisit the topic of the orientation of Etruscan temples starting from Enking's and Pallottino's works that centered on the 16 regions in the Etruscan sky as they were extrapolated from literary sources and from the inscriptions on the well-known Piacenza Liver (Pallottino 1956). Prayon narrowed his focus to 18 structures, not only Etruscan but also from other areas geographically and culturally akin, like Rome, Satricum and Falerii Veteres. In his work, Prayon does not take into consideration structures built after the end of the sixth century BC. This was because of his belief that from that moment on temples as well tombs started to lose their autonomous orientations in favor of the regular orthogonal urbanistic disposition that was making its way through Etruscan cities during the fifth century BC (Prayon 1991: 1286–1287). Prayon inserted these structures, together with the deities they were dedicated to, into a circle divided into 16 equal parts (Fig. 1.1), thus mirroring the 16 regions in the Etruscan sky and inhabited by the deities themselves (Prayon 1991: 1287–1288).

A first result concerned the orientation of those temples towards the southern half of the sky, whereas the worshipped deities were solely present in the northern half. Per Prayon, if a link between temple orientation and celestial dwellings existed, this needed to be investigated in the posterior area of a structure, not in the front. A second result concerned the alignment of all temples dedicated to Uni/Juno in regions 10 and 11, then allowing speculation about a link with the opposite region two in the sky; in the same way, the two temples dedicated to Tinia/Juppiter were both aligned with region seven, therefore linking them with region 15 (Prayon 1991: 1288–1292).

According to Prayon, the link between temple orientation and the celestial dwellings of the worshipped deities needed to be investigated starting from the identified relation between Tinia/Jupiter and region 15 and Uni/Juno and region 2. This is in agreement with what we know from literary sources and the Piacenza Liver[1]: in fact, they assign to Tinia three celestial regions in a row, followed by one single region assigned to Uni. However, compared to most accepted interpretations of the Etruscan sky, this would only work by accepting a shift of such regions two positions down

[1]The reconstruction of the celestial dwellings is based mainly on the elements on the external band of the Piacenza liver (Grenier 1946; Maggiani 1984; Pallottino 1956; Thulin 1906; van der Meer 1987) and on the list of deities described in Mariano Capella's first book (Mar. Cap. I, 41–61; see in particular Weinstock 1946).

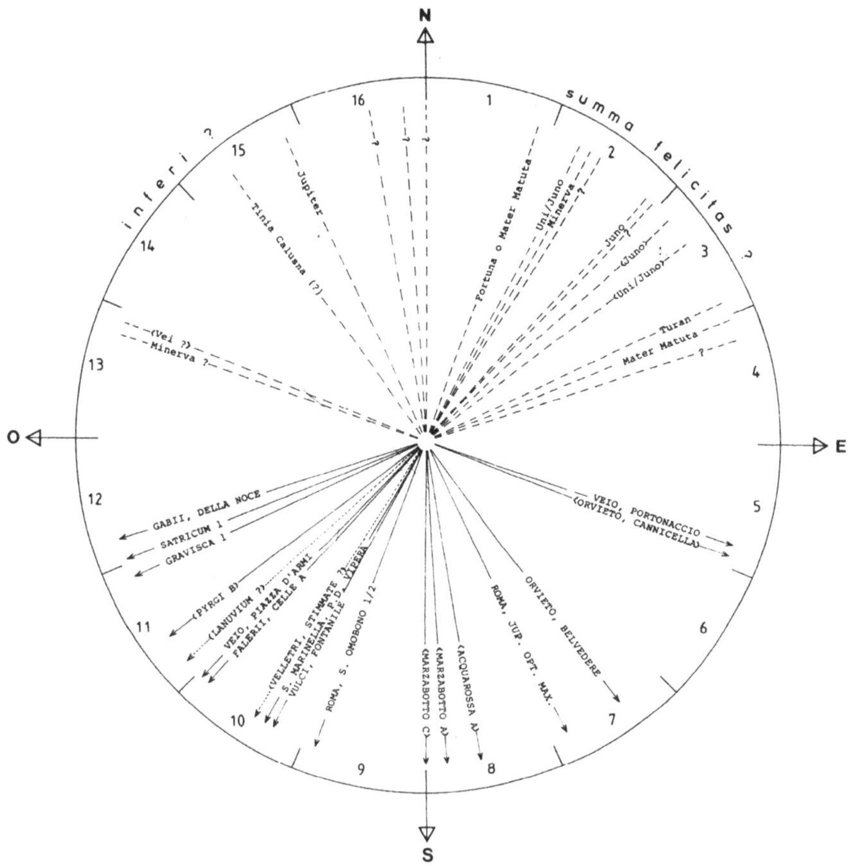

Fig. 1.1 Orientation of Etruscan temples (after Prayon 1991)

towards the west (Prayon 1991: 1293–1294).[2] Per the German scholar's opinion, a shift seems to match the reconstruction of the Etruscan sky offered by van der Meer (1987).

Although they researched the same subject a few years after Prayon, in 1994 Aveni and Romano began their research in a very different way. These scholars started by measuring the azimuth of the temples in the field using a theodolite or a prismatic compass. They studied 32 structures in 23 different locations, and found that many of the published measurements were wrong (Aveni and Romano 1994: 59–62).

Then, Aveni and Romano plotted the measurements that they obtained, placing the azimuth values on the x-axis and the monuments' dates on the y-axis. They came

[2]For the most popular opinion so far among scholars on the temple concept and on the distribution of deities, see Maggiani (1984).

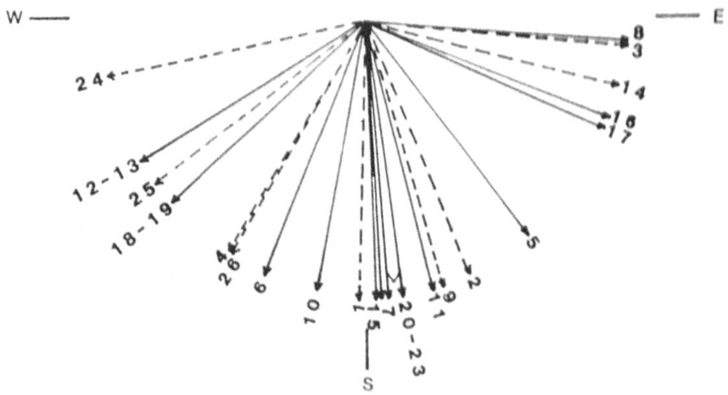

Fig. 1.2 Orientation of Etruscan temples (after Aveni and Romano 1994)

to the conclusion that Etruscan temples were oriented within a sky arc between about 140° and 240°, with the exception of only five structures that had azimuths between 95° and 113° (Fig. 1.2); according to the authors, these maintained traces of an ancient Greek tradition in which temples were positioned towards the sky arc where the Sun rose.[3] Even within the majority group of Etruscan temples Aveni and Romano discerned two different traditions: an older one, dating before the fifth century BC, with solstitial and cardinal orientations, and a more recent one with orientations mostly towards the south (Aveni and Romano 1994: 62).

Although this work was very precise from a methodological point of view (thanks to all the measurements taken in the field and the use of charts), it did not take into account the link between these data and what is known about Etruscan sanctuaries and religion, and therefore left out explanations about the different traditions acknowledged.

Lastly, in 2011 Alfredo Guarino returned to the topic of temple orientation, also dealing with important questions relating to the division of space for divination purposes in Etruscan culture. In his study, 53 edited structures were analyzed, just like Prayon previously did, without using measurements obtained in the field. Among his 53 structures, Guarino recognized 33 that it is safe to consider were temples. He concluded that Etruscan temples were oriented within a sky arc between 108° and 252° (Guarino 2011: 200).

To Guarino (2011: 214–216), this symmetrical orientation to the south had an astronomical explanation since the Sun and the stars rise and set creating angles whose bisector coincides with the south. Thus, Guarino arrived at the hypothesis that a calculation error in the measurement of the equinoxes might be behind this angle, probably due to the wrong usage of the gnomon (Guarino 2011: 216–218). Yet this

[3]The reference is to the Tuscanic temple of Portonaccio in Veio; the sanctuary of Ara della Regina in Tarquinia; the pieve temple in Socana; the one in Fiesole; and the one in Poggio Molina in Populonia (see Aveni and Romano 1994: 65–67, Table 1).

hypothesis is hardly supportable even from a technical point of view and it does not find any confirmation in the archaeological or literary sources where there is no trace of such error; whereas, to the contrary, perfectly perpendicular 'heavenly crosses' are well documented.[4]

Guarino's work certainly increased the number of structures analyzed (even though data were obtained from published layouts and not from direct on-site measurements)[5] and an understanding that orientations were to be investigated not in the celestial dwellings of deities but in the proper sky and in the movements of the Sun and the stars, but his conclusions were difficult to sustain.[6]

The Orientation of Etruscan Temples

Starting from these bibliographical excursions, in the Spring of 2013 I started a series of field campaigns aimed at measuring azimuths and, where it was possible, the horizon heights of the main sacred structures in Etruria. These measurements were carried out using a precision compass and correcting each time the Earth's magnetic field. Regarding the inaccessible structures, I used edited planimetries after comparing them to the ones provided by Google Earth. I analyzed a total of 28 structures that I could be certain were temples, in ten different locations. Nine are Etruscan (Veio, Pyrgi, Vulci, Roselle, Orvieto, Tarquinia, Marzabotto, Volterra, Cerveteri), and one is Faliscan (Falerii Veteres), which is geographically and culturally akin to Etruria.

A comparison between these measurements and the ones in Prayon's, Aveni's and Romano's works (see Table 1.1), not only shows a slightly different record of analyzed structures, but also confirms what has already been speculated before about the non-total accuracy of the azimuths given by the German scholar, mainly due to measurements that were obtained from published layouts and not taken on site, whereas Aveni's and Romano's measurements appear to be more accurate.

The data obtained have been inserted in a chart (Table 1.2), that collects information, where possible, relating to:

---

[4]For specimens with cross signs in the Etruscan world, see Bagnasco Gianni (2008) and Bagnasco Gianni et al. (2016). For literature regarding the concept of *templum* and the division of the sky, see in particular Varr., De ling. lat VII, 6–13; Serv., Ad Aen. II, 693; Serv., Ad Aen. VIII, 427; Serv. Ad Aen. 1, 42; Plin., N. H. II, 137–146; Fest. p. 339L; Plut. Q.R. 78; Cic., De Div. II, 42–43; Cic., De Div. 1, 31; Liv. I, 18. Furthermore, as Guarino himself reminds us, the Greek philosopher Anaximander already came up with the exact calculation of the equinoxes in the fifth century BC (see Franciosi 1990: 81–107).

[5]The result is that, as in Prayon's case, the dataset is not precise and consequently the sky arch obtained for the Etruscan temples orientation will be hard to use, see *infra* (§1.2).

[6]A similar thesis, concerning the seasonal movement of the Sun in relation to the division of the Etruscan sky in 16 regions is applied to the orientation of the sacred architecture of the Etruscans by Stevens (2009). I discuss this thesis in a working paper I recently presented, which is to be approved for publication (Pernigotti 2015–2016).

Table 1.1 A comparison of my measurements and those listed in Aveni and Romano (1994) and Prayon (1991)

Temple	Pernigotti (2018)(°)	Aveni and Romano (1994)(°)	Prayon (1991) (°)
Veio, tuscanic temple of Portonaccio	96	95.5	108
Veio, *oikos* of Piazza d'Armi	217	213	224
Orvieto, Belvedere temple	137	141.4	143
Orvieto, Cannicella	150	ESE	\
Tarquinia, Ara della Regina	95	95	\
Pyrgi, temple A	234	229.8	\
Pyrgi, temple B	232	228	\
Vulci, Fontanile di Legnisina	214	\	209
Vulci, tempio grande	190	192.3	\
Volterra, temples A and B	239.7	239.7	\
Marzabotto, temple A	181	177.8	\
Marzabotto, temple C	181	176.9	\
Falerii, Celle A	154	213	223

1. The orientation of the structures (azimuth, horizon height and declination);
2. Worshipped deities; and
3. Foundation chronology.

The goal was to analyze the orientation of Etruscan temples not only on a general level but also within subgroups, like the one related to the chronological range or to a certain kind of cult.

The data about orientation and worshipped deities have been transferred to a circle to find out about possible links to the sky (Fig. 1.3). It was noticed that most of the temples considered (19 out of 28) were oriented along the sky arc that extended between the points where the Sun rises and sets at the winter solstice. This means that Etruscan temples were oriented not towards sunrise during a particular day of the calendar, as is true for contemporary structures in the Greek world, but so that their frontal faces were touched by the Sun every day for multiple hours a day. On the contrary, the interiors of these structures were never lit by sunlight.

Three small groups were an exception to this rule:

1. A group of five temples oriented towards the east, that were aligned with sunrise during 2 days each year (temples 1, 14, 17, 19, 20);
2. A group of only two temples oriented north-west, whose fronts were never touched by sunlight. They were both connected to catachthonic cults (temples 8, 25); and
3. A group of two temples whose orientations towards sunset on the winter solstice was off the main range by a few degrees (temples 2, 24). To explain these, we can think of a small error in measuring on our part, or by the Etruscan architects.

Table 1.2 Summary table about the orientations of Etruscan temples

	Sacred building	Worshiped deity	Chronology	Bibliography	Azimuth (on field)	Azimuth (from plans)	Horizon height	Declination
Veio								
1	Tuscanic temple (Portonaccio)	Hercle, Rath	500 BC	Colonna (1985a), Colonna (2001)	96°	\	1.9°	−3.177°
2	Sacellum of Menerva (Portonaccio)	Menerva	540/530 BC	Colonna (1985a), Colonna (2001)	244.5°	\	4.6°	−15.374°
3	Oikos (Piazza d'Armi)	Uni?	late VII cent. BC	Bartoloni (2006), Bartoloni et al. (2011)	217°	\	1.5°	−35.141°
Pyrgi								
4	Temple B	Uni/Astarte	510 BC	Colonna (1985b), Colonna (2000)	232°	\	0°	−27.221°
5	Temple A	Thesan—Cavatha?	470 BC	Colonna (1985b), Colonna (2000)	234°	\	0°	−25.223°
6	Sacellum Alpha	Cavatha	c. 350 BC	Colonna (2000), Belelli Marchesini (2013)	228°	\	0°	−29.811°
7	Sacellum Beta	Cavatha/ Demetra—Suri	530 BC	Colonna (2000), Belelli Marchesini (2013)	228°	\	0°	−29.811°
8	Sacellum Gamma	Cavatha—Suri	450 BC	Colonna (2000), Belelli Marchesini (2013)	314°	\	1.64°	32.345°
Vulci								
9	Tempio grande	Menerva	V cent. BC	Colonna and Moretti Sgubini (1985), Moretti Sgubini and Ricciardi (2001)	190°	\	−0.81°	−47.431°
10	Sacellum of Ercole	Ercole	II cent. BC	Colonna (1985c), Moretti Sgubini (2012)	207°/ 208°	\	0.6°	−40.369°
11	Fontanile di Legnisina	Uni	V cent. BC	Massabò and Ricciardi (1988), Moretti Sgubini and Ricciardi (2001)	214°	\	0.95°	−36.923°
12	Carraccio dell'Osteria	Demetra-Vei	V cent. BC	Colonna (1985d), Moretti Sgubini (2012)	150°	\	5.8°	−34.602°

(continued)

Table 1.2 (continued)

	Sacred building	Worshiped deity	Chronology	Bibliography	Azimuth (on field)	Azimuth (from plans)	Horizon height	Declination
Roselle								
13	Temple C	Aiser	VI–V cent. BC	Bartoloni and Bocci Pacini (2002)	200°	\	5.67°	–38.217°
14	Casa con recinto	Divinità femminile	c. 650 BC	Bartoloni and Bocci Pacini (2002)	92°	\	0.95°	–0.821°
Orvieto								
15	Belvedere temple	Tinia Calusna	early V cent. BC	Stopponi (1985a)	137°	\	1.16°	–31.565°
16	Cannicella	Vei—Hercle/ Fauno	late VI cent. BC	Stopponi (1985b), Colonna (1987)	150°	\	0.84°	–38.773°
17	CdF temple A	Tluschva, Kore/ Persefone?, Dioniso?	early IV cent. BC	Stopponi (2012)	\	96°	2.27°	–2.858°
18	CdF temple C	Divinità matronale	late VI cent. BC	Stopponi (2012)	\	218.5°	3°	–32.593°
Tarquinia								
19	Ara della Regina	Hercle?	early VI cent. BC	*Tarchna* IV	95°	\	0.71°	–3.219°
20	Edificio beta	Uni Xia	VII cent. BC	*Tarchna* I	97°	\	1.23°	–4.344°
Marzabotto								
21	Temple of Tinia	Tinia	early V cent. BC	Sassatelli and Govi (2005), Sassatelli (2009)	\	178.5°	10.66°	–35.003°
22	Temple A	\	late VI cent. BC	Vitali (1985)	\	181°	6.37°	–39.302°
23	Temple C	\	late VI cent. BC	Vitali (1985)	\	181°	6.37°	–39.302°

Volterra								
24	Temple B	Papa—Xia	250–200 BC	Bonamici (2003), Bonamici (2005)	\	239.7°	−1.456°	−22.577°
Cerveteri								
25	Vigna Parrochiale (tuscanic temple)	Vei—Tinia	490–480 BC	Maggiani (2001), Bellelli (2008)	317.5°	\	2.49°	36.114°
26	S.Antonio (temple 1)	Hercle	490–480 BC	Maggiani and Rizzo (2001), Maggiani (2008), Rizzo (2008)	206.5°	\	\	\
27	S.Antonio (temple 2)	Rath? Hermes?	490–480 BC	Maggiani and Rizzo (2001), Maggiani (2008), Rizzo (2008)	211.5°	\	\	\
Falerii Veteres								
28	Celle	Juno Curitis	V cent. BC	Colonna et al. (1985)	\	154°	\	\

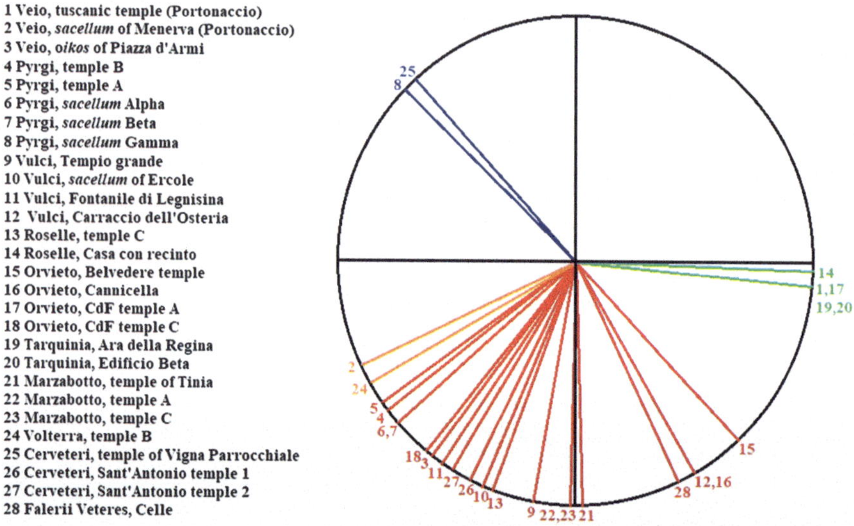

Fig. 1.3 Azimuths of the Etruscan temples measured during the field campaigns of 2013

When comparing these alignments with the worshipped deities, it seems likely that there may have been a link between the orientation of the temples and the associated deities. Thus, we noticed that

1. All temples dedicated to Uni have a very similar orientation (between 214° and 228°);
2. Both temples of Carraccio dell'Osteria and Cannicella, dedicated to Vei, have the same orientation (150°);
3. Temples 1, 17, 19, 20, with identical orientations (95°–96°), show a strong link to Hercle; but this link is attested also in the sacellum of Hercules in Vulci and in the two temples located in Sant'Antonio at Cerveteri, that have an orientation of 207°–208°, almost perpendicular to the previous one; and
4. The case for temples dedicated to Menerva and Tinia is more complicated: they have different orientations but the former are always turned towards the south-west, while the latter are turned towards south-east.

It is harder to find a precise link to the Etruscan celestial sphere as we know it from the Piacenza Liver and the words of Marziano Capella and Plinius, with the 16 celestial dwellings these sources mention (see Notes 1, 2 and 4).

What can be noted though is that.

1. Almost all temples oriented towards the south-western quadrant of the sky, where Etruscans placed terrestrial deities, are dedicated to female deities linked to chthonic cults, with the sole exception of Hercle;
2. Temples 12 and 16, dedicated to Vei and both featuring a strong connection with water cults, are oriented towards the south-east, where deities connected with water reside; and finally

3. Temples 8 and 25, both connected with deities of a catachthonic nature, are oriented towards the north-west, where Etruscans seem to have placed underworld deities.

Concluding Remarks

I believe that the orientation of Etruscan temples was determined by the movement of the Sun, more so than by the celestial dwellings of the deities, with a clear preference for that sky arc where the Sun never rises or sets but where it goes through every day of the year, lighting up the front of the sacred structures for multiple hours a day, for reasons that might be cultual and ritual; this last aspect can certainly be a topic of discussion and in-depth analysis in the future.

At the same time, the two structures Sacellum Gamma in Pyrgi and Vigna Parrocchiale in Cerveteri (tuscanic temple), whose orientation is towards the north-west, seem to show a precise will to turn towards the portion of the sky never affected by the passage of the Sun, with the obvious consequence that their front faces are never illuminated by this star, in full accordance with the underworld cults observed for these temples. Lastly, the aforementioned five temples oriented about towards the east, could have been oriented towards sunrise during a precise day of the year,[7] as seems to happen in Greek tradition.

As far as the individual orientations of various structures and their possible links to the deities worshipped there (and marked by the repetition of similar orientations linked to the same deities), on the one hand we can most likely rule out a connection with the celestial dwellings of the deities the way they are known through the data offered by the Piacenza Liver and by Marziano Capella's words. On the other hand, we have yet to understand the reason for such a possible relation and whether this very link has to be pursued astronomy-wise, connected with the course of the Sun or with one or more stars, or some completely different field.

References

Aveni, A., & Romano, G. (1994). Orientation and Etruscan ritual. *Antiquity, 68*, 545–563.

Bagnasco Gianni, G. (2008). Rappresentazioni dello spazio "sacro" nella documentazione epigrafica Etrusca di epoca orientalizzante. In X. Dupré Raventós, S. Ribichini, & S. Verger (Eds.), *Saturnia Tellus. Definizioni dello Spazio Consacrato in Ambiente Etrusco, Italico, Fenicio-Punico, Iberico e Celtico* (pp. 267–281). Roma: Consiglio Nazionale delle Ricerche.

Bagnasco Gianni, G., Bortolotto, S., & Magli, G. (2013). Astronomy and Etruscan ritual: The case of the Ara della Regina in Tarquinia. *Nexus Network Journal, 15*(3), 1–23.

[7]For the case of Tarquinia see Bagnasco Gianni et al. (2013).

Bagnasco Gianni, G., Gobbi, A., & Scoccimarro, N. (2016). Segni eloquenti in necropoli e abitato. In M. L. Haack (Ed.), *L'Écriture et l'Espace de la Mort. Epigraphie et Nécropoles à l'Époque Préromaine* (pp. 253–302). Roma: Ecole Française de Rome.

Bartoloni, G. (2006). Veio-Piazza d'Armi: Dallo scavo degli Ispettori alle ricerche attuali. In M. Pandolfini Angeletti (Ed.), *Archeologia in Etruria Meridionale. Atti delle Giornate di Studio in Ricordo di Mario Moretti* (pp. 33–47). Roma: L'Erma di Bretschneider.

Bartoloni, G., & Bocci Pacini, P. (2002). Roselle: Una rilettura dei dati di scavo dell'abitato arcaico. In M. Manganelli & E. Pacchiani (Eds.), *Città e Territorio In Etruria. Per una Definizione di Città nell'Etruria Settentrionale* (pp. 187–212). Colle di Val d'Elsa: Grafiche Boccacci.

Bartoloni, G., Acconcia, V., Piergrossi, A., ten Kortenaar, S., & Van Kampen, I. (2011). Veio, Piazza d'Armi: Riconsiderazioni e novità. In *Tetti di Terracotta: La Decorazione Architettonica Fittile tra Etruria e Lazio in età Arcaica* (pp. 116–174). Roma: Officina.

Belelli Marchesini, B. (2013). Le linee di sviluppo topografico del santuario meridionale. In M. P. Baglione & M. D. Gentili (Eds.), *Riflessioni su Pyrgi. Scavi e Ricerche nelle Aree del Santuario* (pp. 11–40). Roma: L'Erma di Bretschneider.

Bellelli, V. (2008). Per una storia del santuario della Vigna Parrocchiale a Cerveteri. In X. Dupré Raventós, S. Ribichini, & S. Verger (Eds.), *Saturnia Tellus: Definizioni dello Spazio Consacrato in Ambiente Erusco, Italico, Fenicio-Punico, Iberico e Celtico* (pp. 319–333). Roma: Consiglio Nazionale delle Ricerche.

Bonamici, M. (2003). *Volterra. L'Acropoli e il Suo Santuario*. Pisa: Giardini editori e stampatori.

Bonamici, M. (2005). Appunti sulle pratiche cultuali nel santuario dell'acropoli volterrana. In M. Bonghi Jovino & F. Chiesa (Eds.), *Offerte dal Regno Vegetale e dal Regno Animale nelle Manifestazioni del Sacro* (pp. 1–10). Roma: L'Erma di Bretschneider.

Colonna, G. (1985a). 5.1 Santuario in località Portonaccio a Veio. In G. Colonna (Ed.), *Santuari d'Etruria* (pp. 99–101). Milano: Electa editrice.

Colonna, G. (1985b). 7.1 Il santuario di Leucotea-Ilizia a Pyrgi. In G. Colonna (Ed.), *Santuari d'Etruria* (pp. 127–130). Milano: Electa editrice.

Colonna, G. (1985c). I santuari urbani. In G. Colonna (Ed.), *Santuari d'Etruria* (pp. 67–69). Milano: Electa editrice.

Colonna, G. (1985d). I santuari suburbani. In G. Colonna (Ed.), *Santuari d'Etruria* (pp. 98–99). Milano: Electa editrice.

Colonna, G. (1987). I culti del santuario della Cannicella. *Annali della Fondazione per il Museo Claudio Faina, 3*, 11–24.

Colonna, G. (2000). Il santuario di Pyrgi dalle origini mitistoriche agli altorilievi frontonali dei Sette e di Leucotea. *Scienze dell'Antichità, 10*, 251–336.

Colonna, G. (2001). Portonaccio. In A. Moretti Sgubini (Ed.), *Veio, Cerveteri, Vulci* (pp. 37–44). Roma: L'Erma di Bretschneider.

Colonna, G., & Moretti Sgubini, A. M. (1985). 4.5 Il tempio grande di Vulci. In G. Colonna (Ed.), *Santuari d'Etruria* (pp. 78–80). Milano: Electa editrice.

Colonna, G., Perina Begni, G., & Melis, F. (1985). 5.2 Il santuario di Celle a Falerii. In G. Colonna (Ed.), *Santuari d'Etruria* (pp. 110–112). Milano: Electa editrice.

Enking, R. (1957). Zur Orientierung der etruskischen Tempel. *Studi Etruschi, 25*, 541–544.

Franciosi, F. (1990). *Le Origini Scientifiche dell'Astronomia Greca*. Roma: L'Erma di Bretschneider.

Grenier, A. (1946). L'orientation du foie de Plaisance. *Latomus, 5*, 293–298.

Guarino, A. (2011). Croce, crux interpretum. Alcune note sulla croce celeste Etrusca, sull'orientamento di templi Etrusco-Italici e sul "fegato di Piacenza". In F. Roncalli (Ed.), *Munuscula. Omaggio degli Allievi Napoletani a Mauro Cristofani* (pp. 183–235). Pozzuoli: Naus.

Maggiani, A. (1984). Qualche osservazione sul fegato di Piacenza. *Studi Etruschi, 50*, 53–88.

Maggiani, A. (2001). L'area della città, La Vigna Parrocchiale. In A. M. Moretti Sgubini (Ed.), *Veio, Cerveteri, Vulci* (pp. 121–122). Roma: L'Erma di Bretschneider.

Maggiani, A. (2008). Il santuario in località S. Antonio a Cerveteri. Il tempio A: La fase Ellenistica. *Mediterranea, V*, 121–137.

Maggiani, A., & Rizzo, M. A. (2001). Area sacra in località S. Antonio. In A. M. Moretti Sgubini (Ed.), *Veio, Cerveteri, Vulci* (pp. 143–145). Roma: L'Erma di Bretschneider.

Massabò, B., & Ricciardi, L. (1988). Vulci, II. Nuove scoperte nel santuario Etrusco di Fontanile di Legnisina. Il tempio, l'altare e il deposito votivo. *Bollettino d'Arte, 73*(48), 27–42.

Moretti Sgubini, A. M. (2012). Vulci. *Bibliografia Topografica della Colonizzazione Greca in Italia e Nelle Isole Tirreniche, 21*, 1082–1154.

Moretti Sgubini, A. M., & Ricciardi, L. (2001). I luoghi di culto. In A. M. Moretti Sgubini (Ed.), *Veio, Cerveteri, Vulci* (pp. 179–180). Roma: L'Erma di Bretschneider.

Pallottino, M. (1956). Deorum sedes. In *Studi in Onore di Aristide Calderini e Roberto Paribeni* (Vol. 3, pp. 223–234). Milano: Ceschina.

Pernigotti, A. (2015–2016). *Materiali per lo Studio del Concetto di Spazio Sacro in Etruria*. Milano: Tesi della Scuola di Specializzazione in Beni Archeologici, Università degli Studi di Milano.

Prayon, F. (1991). *Deorum sedes*. Sull'orientamento dei templi Etrusco-Italici. *Archeologia Classica, 43*, 1285–1295.

Rizzo, M. A. (2008). Scavi e ricerche nell'area sacra di S. Antonio a Cerveteri. *Mediterranea, V*, 91–120.

Sassatelli, G. (2009). Il tempio di Tina a Marzabotto e i culti della città Etrusca. In G. Cresci Marrone & M. Tirelli (Eds.), *Altnoi. Il Santuario Altinate. Strutture del Sacro a Confronto e i Luoghi di Culto Lungo la Via Annia* (pp. 325–344). Roma: Quasar.

Sassatelli, G., & Govi, E. (2005). Il tempio di Tina in area urbana. In G. Sassatelli & E. Govi (Eds.), *Culti, Forma Urbana e Artigianato a Marzabotto. Nuove Prospettive di Ricerca* (pp. 9–62). Bologna: Ante quem.

Stevens, N. L. C. (2009). A new reconstruction of the Etruscan heaven. *American Journal of Archaeology, 113*(2), 153–164.

Stopponi, S. (1985a). 4.7 Il santuario del Belvedere a Orvieto. In G. Colonna (Ed.), *Santuari d'Etruria* (pp. 80–83). Milano: Electa editrice.

Stopponi, S. (1985b). 6.1 Il santuario di Cannicella. In G. Colonna (Ed.), *Santuari d'Etruria* (pp. 116–118). Milano: Electa editrice.

Stopponi, S. (2012). Il Fanum Voltumnae. Dalle divinità Tluschva a San Pietro. *Annali della Fondazione per il Museo Claudio Faina, 19*, 7–75.

Tarchna I. (1997). Bonghi Jovino, M., & Chiaramonte Treré, C. (Eds.), *Tarquinia. Testimonianze Archeologiche e ricostruzione storica. Scavi Sistematici nell'Abitato. Campagne 1982–1988*. Roma: Bretschneider.

Tarchna II. (1999). Chiaramonte Treré, C. (Ed.), *Tarquinia. Testimonianze Archeologiche e Ricostruzione Storica. Scavi Sistematici nell'Abitato. Campagne 1982–1988. I materiali 1*. Roma: Bretschneider.

Tarchna III. (2001). Bonghi Jovino, M. (Ed.), *Tarquinia. Testimonianze Archeologiche e Ricostruzione Storica. Scavi Sistematici nell'Abitato. Campagne 1982–1988. I materiali 2*. Roma: Bretschneider.

Tarchna IV. (2012). Bonghi Jovino, M., & Bagnasco, G.G. (Eds), *Tarquinia. Il Santuario dell'Ara della Regina. I Templi Arcaici*. Roma: Bretschneider.

Thulin, C. (1906). *Die Götter des Martianus Capella und der Bronzeleber von Piacenza*. Gieszen: Alfred Topelmann.

van der Meer, L. B. (1987). *The Bronze Liver of Piacenza. Analysis of a polytheistic sructure*. Amsterdam: Brill.

Vitali, D. (1985). 4.10 L'acropoli di Marzabotto. In G. Colonna (Ed.), *Santuari d'Etruria* (pp. 88–92). Milano: Electa editrice.

Weinstock, S. (1946). Martianus Capella and the cosmic system of the Etruscans. *Journal of Roman Studies, 36*, 101–129.

Chapter 2
Notes on Etruscan Cosmology: The Case of the Tumulus of the Crosses at Cerveteri

Giovanna Bagnasco Gianni

Abstract The aim of this contribution is to test the possibility of the use of cosmological principles connected with Etruscan religion, for composing an inscription which is incised on the wall of a passageway running beneath a ramp attached to the northeast side of the Tumulus of the Crosses, in the Banditaccia necropolis at Cerveteri. The ramp features a stairway leading to a flat ceremonial platform.

On the basis of the letterforms the inscription may be dated to the end of the seventh or the beginning of the sixth century BCE. It is a rare example of a monumental inscription of the Orientalizing period of Etruscan Civilization. Directly beneath the inscription is a sign (*siglum*) formed by a cross inscribed in a circle. This sign has been recognized as the representation of the Etruscan concept of sacred space, whose crucial attributes are delimitation, division and orientation.

A recent new reading of the inscription points out four theonymic elements, which recall divinities that, in the Etruscan cosmology, it may be argued, occupied the northeastern quadrant of the sky.

Any amplification of this recent new reading must take into account interdisciplinary research focused on a possible relationship, in the field of archaeoastronomy, between the theonymic elements and the physical space that they occupy on the wall of the passageway, since the ramp is a crucial element of Etruscan funerary cultic practices.

A Case Study: The Tumulus of the Crosses and Its Inscriptions

The Tumulus of the Crosses (or of the Brooms; see Fig. 2.1) is located in the Banditaccia necropolis at Cerveteri (100 km North of Rome). Its base and the ramp attached to its northeast side were carved into the rock. The ramp is oriented at 22.5° east of north, featuring stairs leading to the top of the Tumulus where cultic

G. Bagnasco Gianni (✉)
Dipartimento di Beni Culturali e Ambientali, Università degli Studi di Milano, Milan, Italy
e-mail: giovanna.bagnasco@unimi.it

G. Magli et al. (eds.), *Archaeoastronomy in the Roman World*, Historical & Cultural Astronomy, https://doi.org/10.1007/978-3-319-97007-3_2

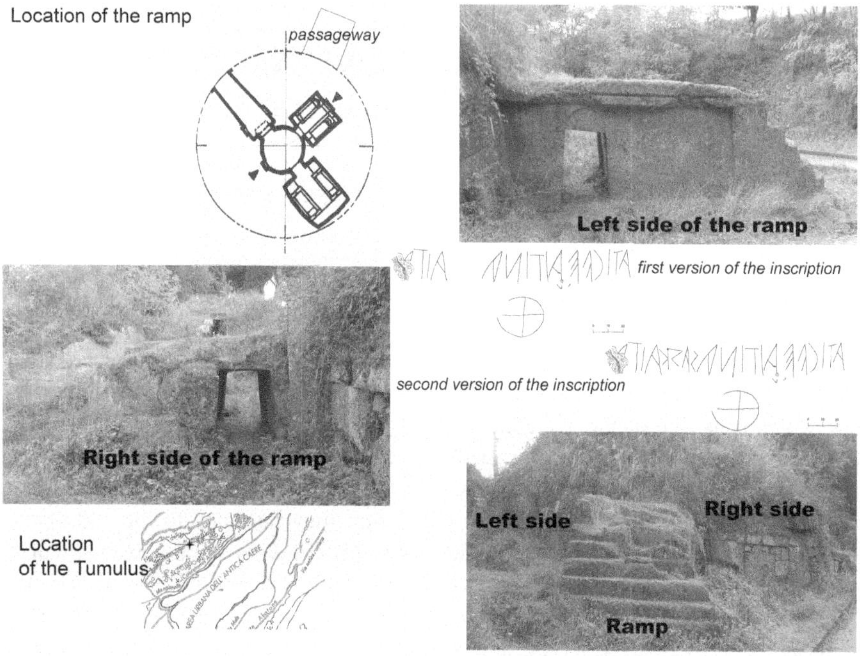

Fig. 2.1 Plans of the Banditaccia necropolis (Cerveteri) and of the Tumulus of the Crosses; views of the Tumulus and of the two sides of the ramp with the passageway; and drawings of the inscription located on the north-east oriented wall of the passageway

practices were celebrated in honour of the predeceased (Fig. 2.1). In Etruscan funerary cultic practices, the ramp is a crucial element, and this is shared with a number of Orientalizing *tumuli* at the same necropolis (Colonna 1986: 398). The ramp, however, is rarely pierced by a passageway, as happens in this case (Bagnasco Gianni 2008).

On the north-east oriented wall of the passageway there is a long inscription reading from right to left (1.48 m long; letters 12–19 cm high); its centre corresponds to that of a sign (or better *siglum*) formed by a cross inscribed in a circle, lacking its upper left arc and located directly beneath the inscription (Fig. 2.1). On the opposite wall an identical sign is reproduced, mirroring the lacking arc of the previous one (Fig. 2.2).

The paleography and layout of both inscriptions (text and *siglum*) suggest that they were engraved together during the late Orientalizing period of the Etruscan Civilization (end of the seventh—beginning of the sixth century BCE) (Bagnasco Gianni 1996: 67–69). Later, but still during the Orientalizing period, four letters were added to the text (Bagnasco Gianni and Facchetti 2015: §1).

It is worth noting that this monumental wall-inscription is the only one known so far.

Fig. 2.2 The passageway under the ramp of the Tumulus with reproductions of the two mirroring *forma quadrans in circulo* located on the two opposite walls

The Epigraphic and Textual Evidence

The most ancient inscription was divided into two parts and cited four theonymic elements, whose linguistic aspects have been explored by Facchetti (Bagnasco Gianni and Facchetti 2015): *atic veai tina* is separated from the last word *aita* by a space (Fig. 2.3).

The suffix *-c* links *Ati*, which means 'mother' in Etruscan, to *Vei*, who is considered a chthonian goddess associated with the generative power of the Earth, counterpart of the Italic goddess of the harvest *Ceres* and of the Greek goddess of fertility Demeter (Bellelli 2012; Simon 2006). The term of kinship with which she is qualified in this inscription underlines her maternal aspect as in the case of other divine entities (Colonna 2012: 206). It is worth noting that this inscription matches Varro's statement about *Ceres*, who also is 'mother' (Rust. 3.1.4): "nec sine causa terram eandem appellabant matrem et Cererem."

Tina, the most important god in the Etruscan pantheon and counterpart of Jupiter/Zeus (Simon 2006: 60), is connected to the daylight, *tin* in Etruscan (Facchetti 2015).

According to the recent reconstruction by Facchetti (Bagnasco Gianni and Facchetti 2015), the second part of the inscription, after the space, cites *Aita*, the god of the underworld and of the night, the counterpart of the Greek Hades (Colonna 2012: 206–207; Krauskopf 2013: 524–525; Rissanen 2012: 129; Simon 2006: 57). Thus, the original version of the inscription, reading from right to left, may be translated as: *Mother Veai Tina*, space, *Aita*.

The second, amplified version of the inscription is obtained by filling the gap of the previous one with four letters, smaller in size: *-sask-* (Fig. 2.4). According to Facchetti (Bagnasco Gianni and Facchetti 2015), they can be split into the derivative morpheme *-asa-* ("what lays beyond") and the suffix *-c* (used in adjectival formations), added to the former *Tina*. The new word *Tinasask* means "the place/land beyond the Light/Tina (Jupiter)", which is suitable for *Aita* and perfectly matches the information given by the first version of the inscription.

Fig. 2.3 Cerveteri, Tumulus of the Crosses: the first version of the inscription

Fig. 2.4 Cerveteri, Tumulus of the Crosses: the second version of the inscription

The meaning of the inscription is now as follows: Mother *Veai* and *Aita* of the "place/land beyond the Light/*Tina*", i.e. "without light" (Bagnasco Gianni and Facchetti 2015).

The *Siglum* Formed by a Cross Inscribed in a Circle

The name of Tumulus of the Crosses was given by Prayon since he recognized in the cross inscribed in the circle the representation of the *Himmelskreuz* (1975: 90), the crucial attributes of which are: orientation, delimitation and division according to Pallottino's definition of the Etruscan concept of sacred space (Bagnasco Gianni 2008). It appears as a mark on a significant number of artifacts, especially vases, used in cultic and funerary practice, disseminated in Etruria over a considerable period of time (Bagnasco Gianni et al. 2016). It has been designated recently as *forma quadrans in circulo* in the signary of the *International Etruscan Sigla Project (IESP)* (Bagnasco Gianni and de Grummond forthcoming).[1]

What Does the Case Study Depict?

The relationship between the inscriptions and the Tumulus can be summarized by means of the following indicators:

1. The *forma quadrans in circulo*, which recalls the shape of the Etruscan sacred space, is inscribed exactly under the centre of the inscription, corresponding to that of the ramp, where the letter '*I*' of *Tina* is inscribed;
2. The missing upper left arc of the *forma quadrans in circulo* is mirrored in the identical sign reproduced on the opposite wall of the passageway;
3. The centre of the ramp is oriented at 22.5° east of north;
4. In the first version of the inscription the name of the god *Aita*, counterpart of the night, is inscribed away from the "Mother *Veai*" and *Tina* by means of a gap;
5. In the second version the location of Aita—away from the "*Mother* Veai", in the "place/land beyond the Light/*Tina*", or "without light/*Tina*"—is specified by means of completion with four more letters (*-sask-*); and
6. The citation of the "Mother *Veai*" is close to *Tina*, the Etruscan counterpart of Jupiter, in both versions of the inscription.

The inscription on the north-east oriented wall of the passageway of the Tumulus is located in the quadrant of the Etruscan cosmos in which Pliny the Elder establishes the most favorable prophecies (*summa felicitas*). Among and in accordance with other literary sources, Pliny describes the *Etruscan* system of dividing the *sky* into a favorable eastern half and an unfavorable western half by means of the north-south axis; a second division generates the basic scheme of four quadrants by means of the east-west axis (*HN* 2.55.142–144); the resulting four quadrants are divided four more times into 16 regions, each of 22.5° (*HN* 2.55.143; Cicero, *De div*. 2.18.42).

The sequence—*Aita*/place (or land) beyond the Light/*Tina*, Mother goddess—matches the beginning of two already known lists of 16 gods reported by two different sources (Table 2.1):

1. The external margin of a bronze stylized sheep's liver, for haruspication,[2] found at Piacenza in 1877 and dated to the first century BCE (Colonna 1984); and
2. The initial regions of a late author's description of the gods invited to the marriage of *Philology* and *Mercury*: Martianus Capella's *De nuptiis Philologiae et Mercurii* (1.45–61) (fifth century CE).

[1] http://www.etruscologia.unimi.it/index.php/progetti/80-progetti/127-etruscanexpo-project

[2] For a recent extensive literary review of haruspication and its formation in Etruria see Bellelli and Mazzi (2013).

Table 2.1 Comparison between the beginning of the lists of 16 gods reported by the Liver and Martianus

Sources	Adjacent positions				
Cells of the liver	*Cilensl*	*Tin Cilen*	*Tin Thuf*	*Tins Then*	*Uni Mae*
Martianus' sequence of regions	*Nocturnus* *Ianitores*	*Jupiter* *Dii Consentes* *Penates* *Salus* *Lares* *Ianus* *Favores opertanei* *Nocturnus*		*Jupiter* *Quirinus Mars* *Lar militaris* *Iuno* *Fons* *Lymphae* *Dii Novensiles*	

The Liver and Martianus' text are our most important sources for the location of the gods according to Etruscan cosmology, but neither source refers to the position of the north.

A number of monuments, mostly stones marked by a cross (*decussis*), show a concrete application on the ground of the division of the Etruscan sky, according to the principles of the *Etrusca Disciplina* (Sassatelli 2017: 188–189; for a different opinion see Maggiani 2009: 236). The Liver is the only evidence we have of an object used for reflecting the cosmos. However, scholars do not unanimously share the same ideas on its orientation (Grenier 1946; Maggiani 1984; Pallottino 1956; Prayon 1991; Stevens 2009; van der Meer 1987).

Since the Tumulus of the Crosses is an oriented monument and the north is well ascertainable, is it possible to use its orientation to find out that of the Liver?

Before trying to answer this question, evidence from literary sources and the bronze liver of Piacenza should be briefly examined as regards the location of *Aita*, counterpart of the Night and of the "place/land beyond the Light/*Tina* (Jupiter)", and of his neighbours.

Connections Between Literary Sources and the Liver

Pliny's and Martianus' division of the sky into 16 regions matches the epigraphic evidence of the 16 cells distributed along the border of the Liver (Colonna 1994; Cherici 2013; de Grummond 2013: 545–546; Krauskopf 2013: 513–514; Maggiani 1984; van der Meer 1987: 22–29). Similarities with the "place/land beyond the Light/*Tina* (Jupiter)" are:

1. The presence of a same divine entity, who is named *Nocturnus* by Martianus and *Cilens* in the Liver, in two adjacent cells (Maggiani 1984: 60, note 34; Pallottino 1956: 225; Torelli 1966: 304–305; van der Meer 1987);

2. The presence of Jupiter in three contiguous regions of Martianus and of *Tina*, his counterpart, in three contiguous cells of the Liver (authors mentioned above); and
3. The quotation of a female goddess after that of *Jupiter Nocturnus* (Martianus)/ *Tina Cilens* (Liver).

Differences Between the Liver, Martianus and Pliny

The main differences concern:

1. The name given to the female goddess located to the left of *Tina*: *Uni Mae* (Liver) and *Juno* (Martianus);
2. The size of the 16 units, which is regular in Pliny (22.5°), irregular in the Liver (de Grummond 2013: 542), and unspecified by Martianus; and
3. The list of gods of the Liver does not exactly match that of Martianus' text and obliges a shift of two regions forward for example in the case of *Uni Mae*/*Juno* and of *Catha*/*Celeritas Solis Filia* (de Grummond 2006: 48–51; Pallottino 1956: 229).

The Name of the Female Goddess and Her Identity

A word inscribed on an Etruscan mirror, *uniapelis*, reflects such a connection between *Uni* and *Mae*, which corresponds to *Maia* and to the name of the month of May, since a gloss (Pallottino 1968: n° 805) explains that the month of May is called *Ampiles* (i.e. *Apelis*) in Etruscan (Briquel 2006: 305; Maras 2009: 308–311; Roncalli 1971–1972: 96–97). Roncalli considers this combination an expression of a particular day devoted to *Uni*. A different interpretation of *Mae* as one of the names of Jupiter is given by van der Meer (1987: 41–43).

However, Macrobius, reporting Cornelius Labeo (*Sat.* I, 12.20–21), explains that *Maia*, due to her *magnitudo*, was also one of the designations of the Earth (... *Maiae, id est terrae* ...) (Torelli 2009: 127) and of a goddess whose name, *Bona Dea*, was secret and unpronounceable (Biondi 2016: 137).

In the sanctuary of Monte Li Santi-Le Rote (Narce, in the Faliscan territory) popular worship addressed Demeter, Persephone, Minerva, Maia and Fortuna from the fifth to the second century BCE. Maia belongs to the later phase and could both depict an aspect of Minerva "*major*" (Biondi 2016: 135–141), but also the above-mentioned prerogatives of the Earth and of Bona Dea (Mastrocinque 2014: 34–35).

Recent evidence from Etruscan sanctuaries shows that from the archaic period *Uni* seems associated with a goddess of the Earth. This happens at other times: at *Veii*, at the sanctuary of Campetti Nord, in which the agrarian aspect of *Vei*, confirmed by a dedication to *Ceres*, is associated with the matronly and protective assignments carried out by *Uni* in the social context (Carosi 2016); at the 'monumental complex' of Tarquinia where, in the archaic period, the chthonic aspect of

Uni (χia) appears in an agrarian setting, which recalls *Ceres* (Bagnasco Gianni 2014); in the votive deposit of the sanctuary of Fontanile di Legnisina at Vulci, where the cult is focused on *Uni* (inscription) and *Demeter* (bronze statuettes) (Ricciardi 2003); and at the northern sanctuary of Pontecagnano (Bailo Modesti et al. 2005: 38), where the chthonic aspect of the main goddess is shared by *Uni* (χia) and a goddess of the earth, **Luas*, a "mother" belonging to Saturn's sphere of influence (Colonna 2009). This association was recently recognized at Marzabotto (Govi 2017: 163–165; Sassatelli 2017: 194, note 25).

It is currently accepted that every pantheon actually is a "stratification" (Devoto 1967: 184), and this is particularly evident in the case of the city of *Veii* homonymous of its poliadic goddess (Bellelli 2012), who was instead identified with *Juno Regina* when *Camillus* took her away from the conquered city in 396 BCE, after her *evocatio* (Colonna 2012: 204, 214; Krauskopf 2013: 514).

The Layout of the Liver

The circumscription of the position of all ancient literary sources and of scholars since the end of the nineteenth century CE, carried out by Pernigotti (2015–2016), points out that scholars still disagree on how it is possible to reconcile the shape of the Liver, the layout of its convex side, and the irregular size of its cells with Pliny's regular scheme of cardinal points. As a matter of fact Pliny's scheme still obliges to look for a sequence of regular cells of the same dimension since Pallottino's analysis (1956: 224) and this is probably one of the reasons why discussion on the orientation of the Liver is ongoing (Gottarelli 2018).[3]

Since we have no evidence to fit the lists of Martianus and the Liver in Pliny's regular scheme, it is worth focusing on the layout of the Liver, which is the only concrete evidence we have so far (Fig. 2.5, with a numbering of the cells according to Maggiani 1984).

Part of the right lobe on the concave side is occupied by the *processus pyramidalis*, the *processus papillaris*, and the *gall-bladder*. The whole surface is divided into 40 cells by engraved lines: 16 run along the margin (1–16); the remaining 24 (17–40) are included in a grid (right lobe, 17–28), in a wheel-like division (left lobe, 31–36) and in the space in between, which corresponds to the direction of the *suspensorium hepatis* on the convex side of the Liver (29–30; 37–40). The *suspensorium hepatis* divides the convex side of the Liver in two lobes: the right lobe is labeled with the name of the Sun (*Usils*) and the other with that of the Moon (*Tiur*).

Cells on the border of the concave side are of unequal size: the smallest are on the right lobe (1–5) where they become increasingly larger (6–8); those on the left lobe

[3]While my paper was already in proofs, I had no time to discuss and integrate recent work by A. Gottarelli (2018).

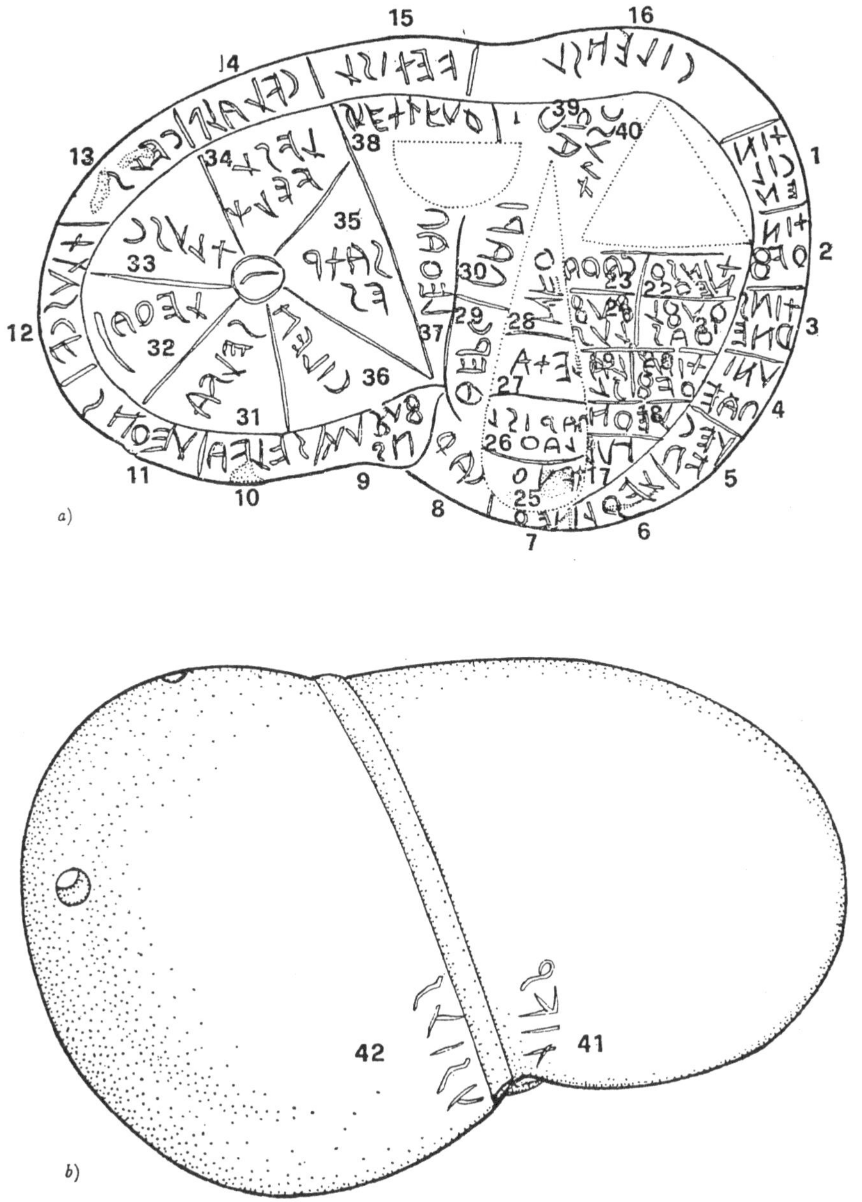

Fig. 2.5 Reproduction of the bronze Liver of Piacenza: (**a**) concave side; (**b**) convex side (after Maggiani 1984: Fig. 1)

are definitively large; the one on the right lobe behind the processus pyramidalis (16) is the largest and has always attracted scholars' attention (Maggiani 1984: 57).

Five cells on the border of the right lobe (1–5) contain two inscribed lines arranged in the *up-down direction; the others* contain a single inscribed line: starting from that behind the *processus pyramidalis,* inscriptions run anticlockwise (12–16) and meet the remaining six, which run clockwise (6–11), near the extreme edge of the left lobe (between 11 and 12).

All inscriptions, except for one of the inner cells (*metlumth*), are written from right to left (Maggiani 1984: 58; van der Meer 1987: 13).

Discussion of the Epigraphic Layout of the Liver

Scholars have explained the *ductus* of the inscriptions as a deliberate choice to make the Liver easily readable and the size of the cells of the Liver as errors of evaluation of space made by the scribe, starting from Pallottino's position in the second part of his contribution (Pallottino 1956: 231–232; cf. Maggiani 1984: 57–58).

According to Maggiani (1984: 57), the engraver started his list of 16 cells from the *incisura umbelicalis* anticlockwise (8–1) and then clockwise (9–16), producing the erroneous disproportion of the last cell (16), due to his miscalculation of the space available. Pallottino's explanation in the first part of his article (1956: 223) is more convincing, given the importance of the bronze Liver for Etruscan cosmology. His idea is that: "…nell'oggetto fosse applicato il sistema dei 'limiti' secondo la disciplina etrusca". As a consequence, the impressive difference in size between cells 16 and 1 could coincide with the engraver's starting point and with the north, since these two adjacent cells also contain the names of two gods (*Cilensl*; *Tin Cilen*), who correspond to the two gods of Martianus' series (*Nocturnus*; Jupiter Nocturnus) (Pallottino 1956: 229). This evidence represents the major difficulty in locating north between cells 2 and 3 according to van der Meer (1987), who follows Deecke's and Thulin's positions.

Maggiani accepted Pallottino's thesis of a cosmological use of the Liver (Maggiani 1984: 68), and the location of the north between cells 16 and 1. However, he noted that, in this case, the engraver could not have started from that point because the strong inclination of the trait separating the two cells does not follow the correct radial orientation centred on the contour line, nor the line established by the base of the *processus pyramidalis*.

Only Colonna (2012: 208) has recently explained the leftwards direction of the gods' names (cells 12–16) through their chthonic nature.

Starting from the assumption that in the Liver haruspication is linked to cosmology (Bellelli and Mazzi 2013: 38; Maggiani 2009: 223), a number of elements noticed by Pallottino and Maggiani could open the way to another perspective on the division of the concave side of the Liver according to a different cosmological ratio:

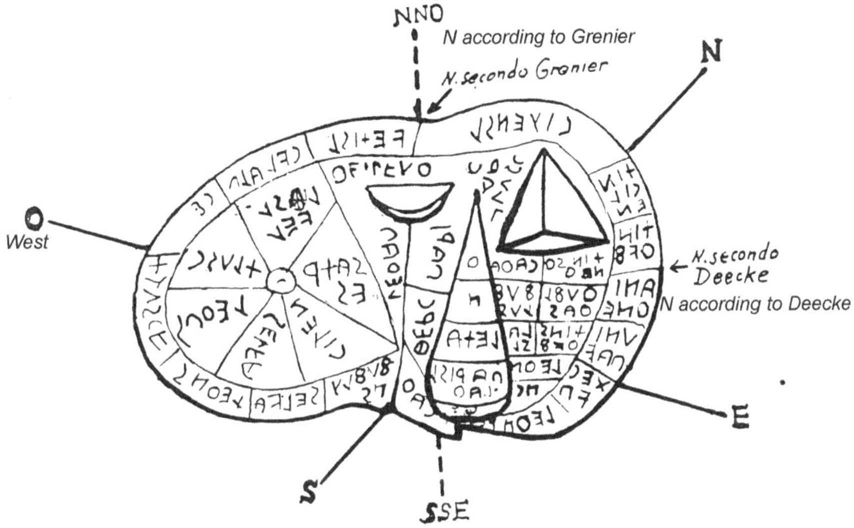

Fig. 2.6 Graphic representation of Pallottino's division of the Liver (unlabeled axes) with the indication of other scholars' theses (after Pallottino 1956: 230)

1. The differences in the sizes and epigraphic layouts of cell 16 and cell 1, which mark the beginning of a change in listing the gods' names down to cell 5 (Pallottino);
2. The peculiar orientation of the trait marking the limit between cell 16 and cell 1 (Maggiani);
3. The north-south axis dividing the surface of the Liver and starting from the limit between cells 1 and 16 does not orthogonally cross the east-west axis; in both cases they do not adhere to the shape of the Liver (Pallottino 1956: 230–232) (Fig. 2.6); and
4. The importance assigned by Varro (Varr. pr. Frontin., in *Corp. Agrim. Rom.* I, p. 27) to the bipartition of the areas of the Moon and the Sun, which should be one of the key points of the Liver as a model for augurs, is underestimated according to Pallottino (1956: 232).

Starting from the evidence that no current theory resolves the problem of the non-orthogonal crossing of two axes, the solution could be to abandon definitively any attempt to fit the cells of the Liver in Pliny's regular scheme and accept the indication given by the epigraphic layout. The purpose could have been to achieve an easier reading of the inscriptions, and also cosmological observations. For example, the effects of the passage of the Sun on the surface of the bronze Liver seems to be depicted on an Etruscan mirror (Fig. 2.7) (Bagnasco Gianni 2012: 300–301). In this case too the inscriptions had to be easily readable.

Following are the epigraphic indicators (Fig. 2.8):

Fig. 2.7 Reproduction of the mirror at the Museo Gregoriano Etrusco with a possible scene of haruspication carried out by the female character in front of the male who is regulating light and shadow with his cloth (Vatican) (after Bagnasco Gianni 2012: 313)

1. On the convex side of the Liver the inscriptions *Usil* and *Tiur* are on the two lobes of the Liver close to the *suspensorium hepatis*;
2. The divergent trait noticed by Maggiani between cells 16 and 1 suggests that one should draw a straight line from this point and reach the opposite edge of the Liver between cells 11 and 12 where the clockwise and anticlockwise *ductus* of the inscriptions meet; and
3. Cells 1–5 containing two lines of inscriptions represent a coherent section; the trait between cells 5 and 6 marks the beginning of another straight line, which separates cells 15 and 14.

Elaborating on the epigraphic indicators:

1. The convex side (Fig. 2.5b): *Usil* and *Tiur* by the *suspensorium hepatis* indicate that this is the north-south axis, according to Grenier's theory (1946); the two areas of the Moon and the Sun respectively coincide with Pliny's main areas (*pars dextera* and *pars sinistra*);
2. The concave side (Fig. 2.9): the two angles produced by the crossing of the two lines mentioned above are regular: two sharp (66°) and two obtuse (114°); and
3. This means that the orientation of the two crossing lines is very near to that of the line of the solstices in Etruria: 60° (sunrise of the summer solstice, 57°); 126° (sunrise of the winter solstice, 123°); 240° (sunset of the winter solstice, 237°); 306° (sunset of the summer solstice, 303°).

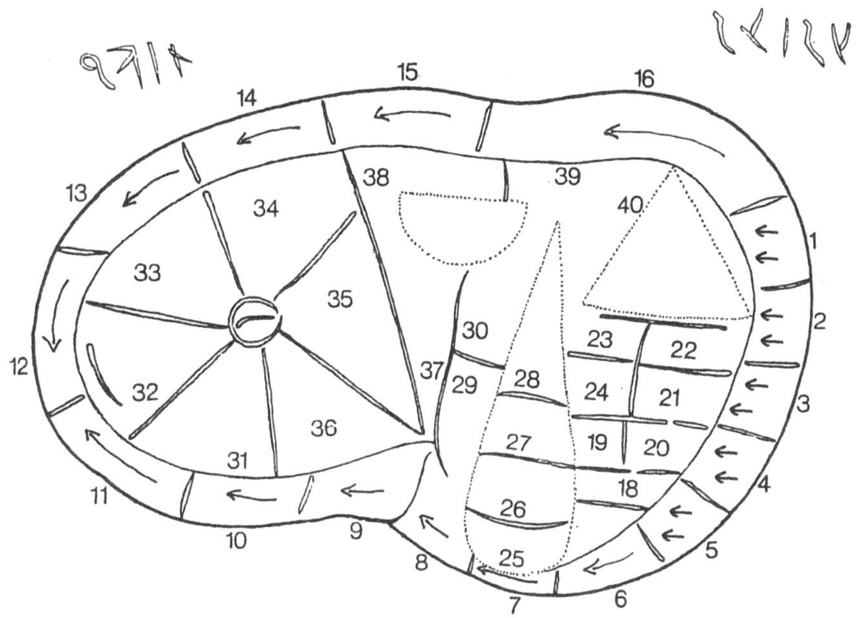

Fig. 2.8 Direction of the inscriptions inside the 16 cells

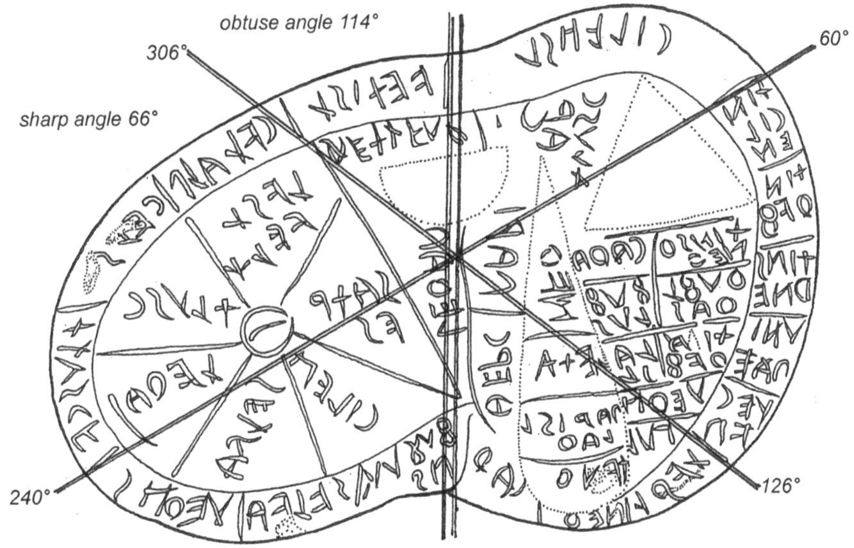

Fig. 2.9 A graphic representation of the projection of the solstices according to the author on Maggiani's drawing of the Liver. The double line in the middle is oriented north-south according to Grenier's theory (1946) and corresponds to the *suspensorium hepatis* on the underside of the Liver, which divides the lobes of the Sun (*Usil*) and the Moon (*Tiur*)

As a consequence of applying the ratio of the solstices, the Liver shows four major coherent areas:

1. with cells 1–5, the Sun rises and lights up the front;
2. with cells 6–11, the Sun never rises and lights up the front; this is the condition of the majority of Etruscan temples (see Pernigotti in this volume);
3. with cells 12–14, the Sun sets and can illuminate while setting; and
4. with cells 15–16, this is really the "place/land beyond the Light" or "without light" inhabited by *Nocturnus*/*Cilens*.

The Evidence of the Tumulus of the Crosses Compared with That of the Literary Sources and the Liver

The boundary between the two regions occupied by the same god, *Nocturnus* in Martianus/*Cilens* in the Liver, has been considered coincident with the north (between regions 16 and 1 (Maggiani 1984; Pallottino 1956). However, as mentioned above (§2.4), neither source refers to the north.

The section of the inscription of the Tumulus containing the indication of the area "beyond the light/*Tina*", where *Aita* stands alone, is oriented to the east of the north since the ramp was located in the north-east quadrant. According to the literary sources (Pliny, Cicero), we are still in the most favorable half of the chart of the sky.

The four theonymic elements of the inscription of the Tumulus appear to correspond to the divine entities quoted in the same order by Martianus and the Liver, even if the Liver is almost five centuries later and Martianus' text almost ten. The profile of the female goddess points out an association between or integration of *Uni* and *Vei*, and the whole group corresponds to other situations in Etruscan sacred areas.

Something similar to the situation depicted by the inscription of the Tumulus is recorded at the sanctuary of the Vigna Parrocchiale at Cerveteri. Here a male god, called *apa* (father), sharing the corresponding infernal and dark prerogatives of Tina/Zeus, Dionysus, *Veiovis*, *Dis Pater* (Colonna 2012: 205–207, 211), is associated with *Vei*/Demeter, while in another area of the same sanctuary an inscription addresses *Uni*/*χia* (Bellelli 2011: 112–114).

In the sanctuary of Volterra there is a similar situation after discovery of a recent fragmentary epigraphic evidence of the name *χia*. The other inscriptions address a male god, called *apa* and *papa* (grandfather), associated with a chthonian goddess named *ati* corresponding to Demeter (Bonamici 2009: 271–272, 275–276; Colonna 2012: 206) in the very context of local ritual practices.

Such correspondences open the way to figure out an encounter between the spheres of competence of *Vei* and *Uni* by means of their chthonian, maternal and terrestrial qualities, which is also shown by the fact that they share the same offering of the piglets (Rafanelli 2013: 574–575). The two goddesses could both be part of the homologation of the maternity cycle with that of wheat, as already pointed out for *Ceres* in the case of the table of Agnone, where she is quoted side by side with a chthonian Jupiter (Prosdocimi 1996: 515–517).

In conclusion, it now seems possible to anchor the beginning of the sequence of the divine entities mentioned on the Liver and by Martianus (see Table 2.1) in the north-east quadrant by means of the orientation of the ramp of the Tumulus: the "place/land beyond the Light/*Tina*" occupies an area east of the north. After space the Light/*Tina* marks his place and, immediately after, there is that of the "Mother".

This order recalls that of the Liver in which the region immediately on the right of the *suspensorium hepatis*, dividing the lobes of the Sun and the Moon on the convex side of the Liver, is occupied by *Cilens* (i.e. the counterpart of *Nocturnus*, according to Martianus) on the concave side, whereas similar dark gods inhabit the westerner regions: *Vetis* (Liver) and *Veiovis* (Martianus). This last correspondence between the two sources matches the above-mentioned nature of darkness shared by *Aita* and *Veiovis* together with the other similar gods identified by Colonna, who also introduces an important difference between chthonian and infernal/dark gods (2012: 208). These last entities occupy Pliny's *regiones maximae dirae* (i.e. the worst regions of the Etruscan cosmos) and match Plautus' description of *Nocturnus* as the god of the starry sky (Amphitryo, 271–284) (Maggiani 1984: 60, note 34; Torelli 1966: 305). Such regions are inhabited by dark gods—*Cilens*/*Nocturnus* and *Vetis*/*Veiovis*—regardless of whether they are males or females, which is one of the most specific aspects of the Etruscan theology (Cristofani 1993).

As a consequence, the cell of the Liver with the names *Tin Cilen*, which follows eastwards the one occupied by *Cilens* alone, where the Sun never rises, could correspond to the beginning of the decreasing of the light (*Tin*) after the sunrise of the summer solstice. From this cell onwards the list of gods of the Liver and that of Martianus hardly coincide, but this is probably due to the absence in the Liver of the main Etruscan goddesses, such as *Vei*, *Menerva* and *Turan*, a problem that should not be underestimated (Colonna 1994: 127). From the point of view of the selection of gods, the Liver could address a local community in a particular situation that has been depicted by Colonna in the framework of the political and historical events of the Hellenistic period (Colonna 1984).

The lacking arc of a circle in the *forma quadrans in circulo* (Fig. 2.2), mirrored by the one on the opposite wall, corresponds to the direction of the axis coming from the north-west and directed inside the Tumulus. The perception of the location of the gods inhabiting the worst regions of the sky could have been reflected in the shape of the *siglum*, intentionally carved without this specific arc of a circle.

Acknowledgment I am grateful to Giulio Magli for the opportunity to discuss this case study, which could open the way to comparing material evidence with a number of intangible aspects described by ancient literary sources concerning Etruscan cosmology, in terms of space and landscape.

References

Bagnasco Gianni, G. (1996). *Oggetti Iscritti di Epoca Orientalizzante in Etruria*. Firenze: Olschki.

Bagnasco Gianni, G. (2008). Rappresentazioni dello spazio "sacro" nella documentazione epigrafica Etrusca di epoca orientalizzante. In X. Dupré Raventós, S. Ribichini, & S. Verger (Eds.), *Saturnia Tellus. Definizioni dello Spazio Consacrato in Ambiente Etrusco, Italico, Fenicio-Punico, Iberico e Celtico* (pp. 267–281). Roma: CNR Edizioni.

Bagnasco Gianni, G. (2012). Tra uomini e dei: funzione e ruolo di alcuni oggetti negli specchi etruschi (con appendici di M. Marzullo, S. Zanni, V. Zenti). In P. Amann (Ed.), *Kulte – Riten – religiose Vorstellungen bei den Etruskern und ihr Verhaltnis zu Politik und Gesellschaft, Akten der 1. Internationalen Tagung der Sektion Wien/Osterreich des Istituto Nazionale di Studi Etruschi ed Italici, Wien 2008* (pp. 287–314). Wien: ÖAW.

Bagnasco Gianni, G. (2014). Una nuova iscrizione dal 'complesso monumentale' della Civita di Tarquinia. In E. Benelli (Ed.), *Cên zic ziχuχe. Per Maristella Pandolfini* (pp. 21–26). Pisa-Roma: Fabrizio Serra Editore.

Bagnasco Gianni, G., & de Grummond, N. T. (forthcoming). *Introducing the International Etruscan Sigla Project*. Accordia Research Institute, University of London.

Bagnasco Gianni, G., & Facchetti, G. M. (2015). Considerazioni sulla rilettura di CIE5992. *Alessandria, 8*(2014), 27–56.

Bagnasco Gianni, G., Gobbi, A., & Scoccimarro, N. (2016). Segni eloquenti in necropoli e abitato. In M. L. Haack (Ed.), *L'Écriture et l'Espace de la Mort. Epigraphie et Nécropoles à l'Époque Préromaine* (pp. 253–302). Roma: Ecole Française de Rome.

Bailo Modesti, G., Frezza, A., & Lupia, A. (2005). Le acque intorno agli dei. Rituali e offerte votive nel santuario settentrionale di Pontecagnano. In M. Bonghi Jovino & F. Chiesa (Eds.), *Offerte dal Regno Vegetale e dal Regno Animale nelle Manifestazioni del Sacro Atti dell'Incontro di Studio (Milano 26–27 Giugno 2003)* (pp. 37–63). Roma: L'Erma di Bretschneider.

Bellelli, V. (2011). Un'iscrizione greca dipinta e i culti della Vigna Parrocchiale a Caere. *Studi Etruschi, 74*(2008), 91–124.

Bellelli, V. (2012). Vei: Nome, competenze e particolarità cultuali di una divinità etrusca. In V. Nizzo & L. La Rocca (Eds.), *Antropologia e Archeologia a Confronto: Rappresentazioni e Pratiche del Sacro. Atti dell'Incontro Internazionale di Studi (Roma, 20–21 Maggio 2011)* (pp. 455–478). Roma: ESS Editorial Service System.

Bellelli, V., & Mazzi, M. (2013). *Extispicio. Una Scienza Divinatoria tra Mesopotamia ed Etruria*. Roma: Scienze e Lettere.

Biondi, L. (2016). Divinità e culti nel santuario di Mazzano: Evidenze linguistiche. In M. A. De Lucia Brolli (Ed.), *Il Santuario di Monte Li Santi – Le Rote a Narce. Scavi 1985–1996. Parte I. La Topografia, le Fasi, il Culto* (pp. 128–148). Pisa-Roma: Fabrizio Serra Editore.

Bonamici, M. (2009). *Volaterrae*. Piano di Castello. Santuario dell'acropoli. *Studi Etruschi, 73* (2007), 268–293.

Briquel, D. (2006). Les gloses étrusques. *Res Antiquae, 3*, 301–318.

Carosi, S. (2016). Il santuario di Campetti Nord a Veio. Alcune riflessioni sul regime delle offerte e sulla diffusione di elementi demetriaci nel culto. In A. Russo & F. Guarneri (Eds.), *Santuari Mediterranei tra Oriente e Occidente. Interazioni e Contatti Culturali* (pp. 215–218). Scienze e Lettere: Roma.

Cherici, A. (2013). The science of the Etruscans. In J. MacIntosh Turfa (Ed.), *The Etruscan world* (pp. 683–694). London and New York: Routledge.

Colonna, G. (1984). Il Fegato di Piacenza e la tarda Etruscità cispadana. In P. Delbianco (Ed.), *Culture Figurative e Materiali tra Emilia e Marche. Studi in Memoria di Mario Zuffa* (pp. 171–180). Rimini: Maggioli.

Colonna, G. (1986). Urbanistica e Architettura. In G. Pugliese Carratelli (Ed.), *Rasenna. Storia e Civiltà degli Etruschi* (pp. 369–530). Milano: Libri Scheiwiller.

Colonna, G. (1994). A proposito degli dei del Fegato di Piacenza. *Studi Etruschi, 59*(1993), 123–139.

Colonna, G. (2009). *Picentia*. Pontecagnano. *Studi Etruschi, 73*(2007), 359–361.

Colonna, G. (2012). I santuari comunitari e il culto delle divinità catactonie in Etruria. *Annali della Fondazione per il Museo "Claudio Faina", XIX*, 203–226.

Cristofani, M. (1993). Sul processo di antropomorfizzazione nel pantheon etrusco. In *Miscellanea Etrusco-Italica* (Vol. 1, pp. 9–21). Roma: CNR Edizioni.
de Grummond, N. T. (2006). *Etruscan myth, sacred history, and legend*. Philadelphia: University of Pennsylvania Museum of Archaeology and Anthropology.
de Grummond, N. T. (2013). Haruspicy and augury: Sources and procedures. In J. MacIntosh Turfa (Ed.), *The Etruscan world* (pp. 539–556). London and New York: Routledge.
Devoto, G. (1967). Il panteon di Agnone. *Studi Etruschi, 25*, 179–197.
Facchetti, G. M. (2015). Tinas cliniiaras. *AION Linguistica, 4*, 135–176.
Gottarelli, A. (2018). *Padānu. Un'ombra tra le mani del tempo. La decifrazione funzionale del fegato etrusco di Piacenza (Archeologia del Rito, 3)*. Bologna: Edizioni Te.m.p.l.a.
Govi, E. (2017). La dimensione del sacro nella città di *Kainua*-Marzabotto. In E. Govi (Ed.), *La Città Etrusca e il Sacro. Santuari e Istituzioni Politiche. Atti del Convegno (Bologna 21–23 Gennaio 2016)* (pp. 145–179). Bologna: Bononia University Press.
Grenier, A. (1946). L'orientation du foie de Plaisance. *Latomus, 5*, 293–298.
Krauskopf, I. (2013). Gods and demons in the Etruscan pantheon. In J. MacIntosh Turfa (Ed.), *The Etruscan world* (pp. 513–527). London and New York: Routledge.
Maggiani, A. (1984). Qualche osservazione sul fegato di Piacenza. *Studi Etruschi, 50*, 53–88.
Maggiani, A. (2009). *Deorum sedes*: divinazione Etrusca o dottrina augurale Romana? *Annali della Fondazione per il Museo "Claudio Faina", XVI*, 221–237.
Maras, D. F. (2009). *Il Dono Votivo. Gli dei e il Sacro nelle Iscrizioni Etrusche di Culto (Biblioteca di Studi Etruschi 46)*. Pisa-Roma: Fabrizio Serra Editore.
Mastrocinque, A. (2014). *Bona Dea and the Cults of Roman Women (Potsdamer Altertums-wissenschaftliche Beiträge, 49)*. Stuttgart: Franz Steiner Verlag.
Pallottino, M. (1956). Deorum sedes. In *Studi in Onore di Aristide Calderini e Roberto Paribeni* (Vol. 3, pp. 223–234). Milano: Ceschina.
Pallottino, M. (1968). *Testimonia Linguae Etruscae*. Firenze: Olschki.
Pernigotti, A. (2015–2016). *Materiali per lo Studio del Concetto di Spazio Sacro in Etruria*. Milano: Tesi della Scuola di Specializzazione in Beni Archeologici, Università degli Studi di Milano.
Prayon, F. (1975). *Frühetruskische Grab- und Hausarchitektur*. Heidelberg: F.H. Kerle.
Prayon, F. (1991). Deorum sedes. Sull'orientamento dei templi Etrusco-Italici. *Archeologia Classica, 43*, 1285–1295.
Prosdocimi, A. L. (1996). La tavola di Agnone. Una interpretazione. In L. Del Tutto Palma (Ed.), *La Tavola di Agnone nel Contesto Italico. Atti del Convegno di Studio (Agnone, 13–15 Aprile 1994)* (pp. 435–630). Firenze: Olschki.
Rafanelli, S. (2013). Haruspicy and augury: Sources and procedures. In J. MacIntosh Turfa (Ed.), *The Etruscan World* (pp. 566–593). London and New York: Routledge.
Ricciardi, L. (2003). Deposito votivo del santuario di Fontanile di Legnisina a Vulci. In D. Zinelli (Ed.), *L'Acqua degli dei: Immagini di Fontane, Vasellame, Culti Salutari e in Grotta* (pp. 125–131). Montepulciano: Le Balze.
Rissanen, M. (2012). The "Hirpi Sorani" and the wolf cults of Central Italy. *Arctos. Acta Philologica Fennica, 46*, 115–135.
Roncalli, F. (1971–1972). Uno *specchio* del Museo Gregoriano con iscrizione etrusca inedita. *Rendiconti. Pontificia Accademia Romana di Archeologia, XLIV*, 75–97.
Sassatelli, G. (2017). La città e il sacro in Etruria padana: Riti di fondazione, culti e assetti urbanistico-istituzionali. In E. Govi (Ed.), *La Città Etrusca e il Sacro. Santuari e Istituzioni Politiche. Atti del Convegno (Bologna 21–23 Gennaio 2016)* (pp. 181–204). Bologna: Bononia University Press.
Simon, E. (2006). Gods in harmony: The Etruscan pantheon. In N. Thomson de Grummond & E. Simon (Eds.), *The religion of the Etruscans* (pp. 45–65). Austin: University of Texas Press.
Stevens, N. L. C. (2009). A new reconstruction of the Etruscan heaven. *American Journal of Archaeology, 113*(2), 153–164.

Torelli, M. (1966). Un templum augurale d'età repubblicana a Bantia. *Rendiconti dell'Accademia dei Lincei, XXI*, 293–315.
Torelli, M. (2009). Religione e rituali dal mondo latino a quello etrusco: Un capitolo della protostoria. *Annali della Fondazione per il Museo "Claudio Faina", XVI*, 119–154.
van der Meer, L. B. (1987). *The Bronze Liver of Piacenza. Analysis of a polytheistic structure.* Amsterdam: Brill.

Part II
Rome

Chapter 3
The Orchestration of Time in Ancient and Medieval Buildings

Robert Hannah

Abstract According to the Neoplatonist Porphyry, the followers of Mithras regarded the midsummer solstice as the point where souls descended in *genesis* from the heavens into this world, while at the midwinter solstice they re-entered the heavens in *apogenesis*. The Sun-god Mithras oversaw this migration of souls from his seat at the equinoxes, midway between the solstices. Beyond Mithraism Porphyry attributes this notion of psychic entry and exit through the solstitial points to his Platonic predecessors, Numenios and Kronios, and in some respects even back to Plato himself. This chapter examines to what extent such a philosophy or theology is discoverable also in Imperial and Late Antique Roman buildings, notably in the domed spaces of Nero's Domus Aurea, Domitian's Palace, the Pantheon and (their architectural descendants) the churches of Byzantium. It is found that there are grounds for seeing the pagan Imperial buildings as designed around lighting effects at the times of the equinoxes and solstices, perhaps to enable imperial hierophanies. Perhaps reflecting this aspect of Roman domed structures, a divine hierophany is undoubtedly intended to be highlighted in the later Byzantine churches, which were designed to allow a shaft of light to fall on the altar through the apse windows at those moments in the liturgy when the Holy Spirit, represented by the beam of light, was imagined as descending to sanctify the offerings presented to God by the priest in front of the altar. These architecturally orchestrated effects could be particularly significant around the times of the winter solstice and the vernal equinox, which were associated liturgically with Christmas and Easter respectively.

Introduction

In this paper I wish to examine metaphysical conceptions of time that appear to be bound up within some of the major public architectural projects of the early Roman Empire and their descendants in the Byzantine Empire. My motivating hypothesis is that in both public and private, élite and plebeian contexts, there was a belief that the

---

R. Hannah (✉)
University of Waikato, Hamilton, New Zealand

G. Magli et al. (eds.), *Archaeoastronomy in the Roman World*, Historical & Cultural Astronomy, https://doi.org/10.1007/978-3-319-97007-3_3

pivotal times of the equinoxes in the solar year were points of balance, where the divinity, or the deified emperor, would sit, while the solstices north and south of these mid-points provided 'passageways' between the time-bound, mortal world of the 'here and now' and the timeless, immortal world of the 'before' and 'after'.

My hypothesis stems from an allegorical interpretation by the third century CE Neoplatonic philosopher, Porphyry, of a description of a cave sacred to the nymphs in Homer's *Odyssey* (*Od.* 13.102–112). In this commentary Porphyry equates the cave with the cosmos, a conception which goes back to Plato himself. But Porphyry also draws on beliefs attributed to the Mysteries of Mithras, which was a major cult of the Empire from the late first century CE. In both Mithraic and Platonic circles the cave-cosmos had a point of entry for souls coming into being in this world and a point of exit for souls leaving it. I will argue that this conception of the cosmos pervaded Roman society beyond these philosophical and cultic instances, and is probably realised in public architectural projects which made use of the dome, which was itself equated with the heavens. Such an interpretation is most easily made initially for the Pantheon in Rome. An antecedent for this may be found in the palace of Nero, the Domus Aurea. But do other projects lie between these two buildings to provide some continuity of form and concept? And does the concept continue beyond the time of the Pantheon? For this extension of my investigation, I will look first at the Palace of Domitian in Rome, which lies in time between the Domus Aurea and the Pantheon; and then much later at medieval descendants of the Pantheon—churches of the Byzantine world. These later buildings may then help us to see how the interaction between people and light was orchestrated at special times of the year, which is an aspect of the lived experience of the imperial structures that we lack in the literary and archaeological sources.

The Mysteries of Mithras

In the *Odyssey*, Odysseus was stranded on the island of the Phaiakians while on his way back from the war at Troy to his home on the island of Ithaka. At the start of Book 13, the Phaiakians take Odysseus back to Ithaka:

> At the head of the harbour is a long-leafed olive tree, and near it a pleasant, shadowy cave sacred to the nymphs that are called Naiads. Therein are mixing bowls and jars of stone, and there too the bees store honey. And in the cave are long looms of stone, at which the nymphs weave webs of purple dye, a wonder to behold; and therein are also ever-flowing springs. Two doors there are to the cave, one toward the North Wind, by which men go down, but that toward the South Wind is sacred, nor do men enter thereby; it is the way of the immortals. (Homer, *Odyssey* 13.93–112).[1]

[1] All translations are my own.

In the course of an allegorical interpretation of the significance of the cave,[2] the Neoplatonic philosopher Porphyry (CE 234–305) reports:

> Thus also the Persians say with respect to the descent of the souls below and the way out again, naming the place a cave when they introduce the initiate to the rites. As Euboulos says, Zoroaster was the first to dedicate a natural cave, with flowers and streams, in the neighbouring mountains of Persia, in honour of Mithras, the maker and father of all, a cave bearing a likeness to the cosmos that Mithras had created, and the things inside bore symbols of the cosmos's elements and regions at commensurate distances . . . (Porphyry, *On the Cave of the Nymphs* 6).

So the cult centre, the cave, had cosmological significance. It symbolised the cosmos itself, and it was dedicated in honour of Mithras, 'the maker and father of all'. After Zoroaster others then followed suit, with the Pythagoreans and Plato also envisaging the cosmos as a cave (*On the Cave of the Nymphs* 8).

For Porphyry the cave is associated also with the element of moisture and through that with the nymphs who preside over streams and springs. The watery element is important because souls that are descending into *genesis*, i.e. passing into life in this world, were thought to settle by water (Porphyry, *On the Cave of the Nymphs* 13–20).

Porphyry then proceeds to explore Homer's cave. It has two entrances, one facing north and used by mortals, the other south and used by immortals (*On the Cave of the Nymphs* 20). The philosopher explains:

> The cave being a likeness and symbol of the cosmos, Noumenios and his pupil Kronios say there are two extremities in heaven. Nothing is more southerly than the winter one and nothing more northerly than the summer one. The summer one is in Cancer, while the winter one is in Capricorn. (Porphyry, *On the Cave of the Nymphs* 21).

'Nothing' in this passage must mean 'nothing on the solar and planetary path'. It is particularly the Sun's apparent path across the sky that interests Porphyry. At Mediterranean latitudes the Sun traverses through the year a limited arc along the horizon. The Sun will neither rise nor set outside the arc on the eastern and western horizons between the two solstices, the winter solstice in late December when the Sun is in Capricorn, and the summer solstice in late June, when it is in Cancer.

Porphyry continues, drawing on his Platonist predecessors, Numenios and Kronios (mid-second century CE):

> The theologians therefore make these two gates, Cancer and Capricorn, but Plato says they are two openings. Of these they say Cancer is the gate through which the souls descend, and Capricorn the one through which they ascend. Cancer is northern and enables descent, while Capricorn is southern and enables ascent. The northern belongs to souls descending into generation (*genesis*). And the gates of the cave that face to the north are rightly for the descent for men; but the southern are not for the gods, but for those ascending to the gods . . . (Porphyry, *On the Cave of the Nymphs* 22–23).

[2]For a fuller analysis of Porphyry's commentary on the Homeric passage, see Hannah (2019).

The midsummer and midwinter solstitial points are significant because they are the extreme points of the Sun's apparent path across the heavens. These points represent gateways for entry into our world and then exit from it.[3]

Porphyry concludes that:

> Therefore he [Homer] dedicated the gates neither to the east and west nor to the equinoxes, namely Aries and Libra, but to the south and north, and to the gates most southerly to the south and those most northerly to the north, because he dedicated the cave to souls and water nymphs. And for souls these places are fit for generation (*genesis*) and return from generation (*apogenesis*). (Porphyry, *On the Cave of the Nymphs* 24).

At this point Porphyry refers this vision of the cosmos again to the beliefs of the cult of Mithras:

> Therefore they assigned to Mithras a seat at the equinoxes. For this reason he carries the sword of Aries, the sign of Ares, and rides a bull, as Taurus belongs to Aphrodite. Being creator and lord of generation (*genesis*) Mithras is appointed over the equinoctial circle, with the north on his right and south on his left, and Cautes is appointed to the south because it is hot, and <Cautopates> to the north because of the cold of the wind. (Porphyry, *On the Cave of the Nymphs* 24).

Here we may usefully refer the interpretation of the cave to the central visual icon of the cult, the image of Mithras stabbing the bull (Fig. 3.1). Mithras, the creator and lord of *genesis* is in the centre, astride the Bull, at the equinox, which is in Aries. Cautes stands on the viewer's left with a torch upraised, while Cautopates stands on the right with a downturned torch. These two figures stand for the solstices, Cautes for the winter at the tropic of Capricorn in the south, Cautopates for the summer at the tropic of Cancer in the north. The solstices are 'passageways' for souls descending in *genesis* to the mortal world, and ascending in *apogenesis* to the immortal world (see Beck 2006; Hannah 1996).

As we have seen, Porphyry relies at certain points in his interpretation on his Platonic predecessors, Numenios and Kronios. But he also feels able to take it back to Plato himself, who, he reminds his readers, "... says that there are two openings, one through which souls ascend to the heavens, the other through which they descend to earth." (*On the Cave of the Nymphs* 29, referring to Plato, *Laws* X 896e5–6).

To recap: according to Porphyry, in Mithraic belief, the midsummer solstice was regarded as the point of entry for souls from the heavens into this world, enabling *genesis*. Here the Sun was in Cancer in June and at its most northerly, and suited for

[3]At *On the Cave of the Nymphs* 28.1–7 Porphyry adds that 'Capricorn and Cancer are near the Milky Way, having been assigned its limits, Cancer its northern, and Capricorn its southern. According to Pythagoras the people of dreams are the souls which they say are gathered in the Milky Way, which is so called from their being nursed on milk, whenever they fall into *genesis*.' So here the souls are envisaged as occupying the Milky Way, which they exit from and re-enter at the supposed points of intersection with the zodiac at Cancer and Capricorn respectively. The actual points of intersection, however, lie a sign away, at Gemini and Sagittarius. This error is repeated in the early fifth century by Macrobius, in his *Commentary on the Dream of Scipio* 12; see Stahl 1952: 133–134.

Fig. 3.1 Roman relief sculpture of Mithras, from the Walbrook Mithraeum. London, Museum of London, inv. no. A16933. Left: Cautes/South/Winter/Capricorn/Ascent/*Apogenesis*. Centre: Mithras/Equinox/Aries. Right: Cautopates/North/Summer/Cancer/Descent/*Genesis* (image © Museum of London)

descent into this world. At the midwinter solstice, on the other hand, there lay the point of re-entry to the heavens, which was called *apogenesis*, the return from *genesis*. Here the Sun was in Capricorn in December and at its most southerly, and suited to ascent into the upper world. In the Mithraic cult, the Sun-god Mithras oversaw this migration of souls from his seat at the equinoxes, midway between the solstices. The cult's meetings and rituals took place in a cave-like setting, whose form was regarded as a symbol of the cosmos.

Is such astronomical symbolism recognisable elsewhere, especially in public contexts?

Rome: The Pantheon

The Pantheon was originally built by Agrippa at the southern end of the Campus Martius around 27 BCE, probably as a circular building more or less open to the sky and oriented to the north. With the Mausoleum of Augustus at the northern end of the Campus Martius it helped to frame a distinctly Augustan space in this part of Rome (see Lange 2015: 142, and 286 n.60 for references to the question of its orientation). This structure was destroyed by fire in CE 80 and restored under Domitian, then burned again and rebuilt from the time of Trajan until it was completed in its present form under Hadrian in ca. CE 128 (Hannah and Magli 2011).

The historian Cassius Dio, writing some 70 years after Hadrian, indicates that the meaning of the building's name was problematic in his own time:

> Perhaps it has this name because, among the statues which embellished it, there were those of many gods, including Mars and Venus; but my own opinion on the origin of the name is that, because of its vaulted roof, it actually resembles the heavens. (Cassius Dio 53.27.2).

In general, the building from Agrippa to Hadrian may fairly be regarded as a monument to the Julian family, to which Augustus belonged (Beaujeu 1955: 122–123; Coarelli 1983: 42; Scheid 1990: 461). Apart from statues of the gods connected with the family that are mentioned by Dio, there were figures of Augustus himself and Agrippa in the porch. Also according to Dio (53.27.3) Agrippa's original intention had been that the building should be named after Augustus, presumably as an 'Augusteum', but this move was rejected by the Emperor.

The Pantheon certainly had religious functions—the images of the gods would suggest as much. In addition, the Arval Brethren, an ancient college of 12 priests that was concerned with ensuring agricultural fertility (Varro, *On the Latin Language* 5.85), and which was revived by Augustus, met at the Pantheon in Nero's reign in January CE 59, and presumably in previous years too from the time of Augustus. The Emperor himself was a member of the priesthood. At the Pantheon they performed the ritual (*indictio*) of proclaiming the coming festival of Dea Dia, an agrarian goddess. From CE 63 they conducted this ritual instead at the Temple of Concord at the northern end of the Roman Forum. Records of the proclamation indicate that the proclaimer was to stand "... with washed hands, with veiled head, in the open air, beneath the gable, facing east ..." (Beard 1985: 152), regardless of whether the site was the Pantheon or the Temple of Concord. Sperling (2004: 67) has attractively suggested that as the cult "... involved the calculation of times for the planting and harvesting of crops, acts which at that time, as today, could be related to the religious calendar ...", then it raised questions of astronomical concern, and that these were considered and calculated in the Pantheon. The goddess's festival was usually in May and the grove sacred to her was situated just outside Rome (Beard 1985: 130, 133, 149–156; Iara 2015: 128–130; Scheid 1990: 176–178, 460–464, 708–732).

But the Pantheon could also be used for secular functions. Dio (69.7.1) relates how Hadrian

> ... did all the important and most necessary business through the senate, and he sat in judgement with the top men, sometimes in the Palace, sometimes in the Forum or the Pantheon or in many other places, from a tribunal, so that whatever was done was in the public eye.

The Pantheon as rebuilt by Trajan and Hadrian faces north. It comprises a rectangular columned portico in front of a rotunda roofed by a huge hemispherical dome 43.3 m in diameter, which is itself built over a cylinder that has the same diameter and is as high as half of it. The roof is punctured by an oculus 8.3 m wide, which is the only source of direct light, since no direct sunlight can enter from the

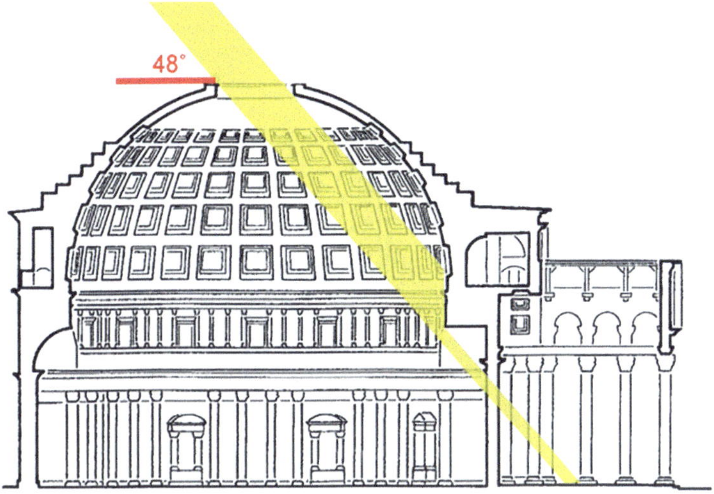

Fig. 3.2 Rome, the Pantheon. The shaded area shows the sunlight through the oculus at noon on the equinoxes, when the Sun's altitude is 48° (after Hannah and Magli 2011: 493, Fig. 4)

northern door throughout the whole year. The ceiling looks like a perfect hemisphere, but in structural terms this is an illusion: as can be seen from the outside, the dome is not hemispherical but flattened, and the appearance of a hemisphere on the interior is due to non-structural infilling.

The interior of the building, with its limited lighting, resembles a particular type of sundial, called the *hemicyclium* (Hannah 2009a: 145–154; 2009b; Potamianos 1996: 105). This consisted of a stone block carved out into a hollow hemisphere, with a hole let into its upper surface, through which the sunlight filtered on to the south-facing surface inside, where a series of reference lines was incised as in standard sundials, typically to signify no more than the summer and winter solstices and the spring and autumn equinoxes (the last as a single line between the outer two), in addition to the daily hour-lines.

We can see the same effect of sunlight within the Pantheon as we see in the *hemicyclium*, although here we must focus only on the effect at noon, which lies pretty much on the axis of the building. Beyond that moment in the day the inner decoration of the ceiling, with its horizontal bands of coffers, cannot work exactly as a sundial, whose markings would instead lie on parabolic curves.

At noon we are halfway through the daylight period. At the equinoxes, we are halfway through the Sun's annual cycle between one solstice and the other. It is remarkable that at this moment the Sun strikes the interior of the Pantheon over the entrance doorway precisely at the juncture between the base cylinder and the roof's dome, which lies halfway up the height of the building (Fig. 3.2).

So halfway up the building's height, where dome and cylinder meet, and halfway in the Sun's cycle in the year and day, time and space are coordinated. That this is by design, not due to coincidence, is indicated by the artificial nature of the interior hemisphere.

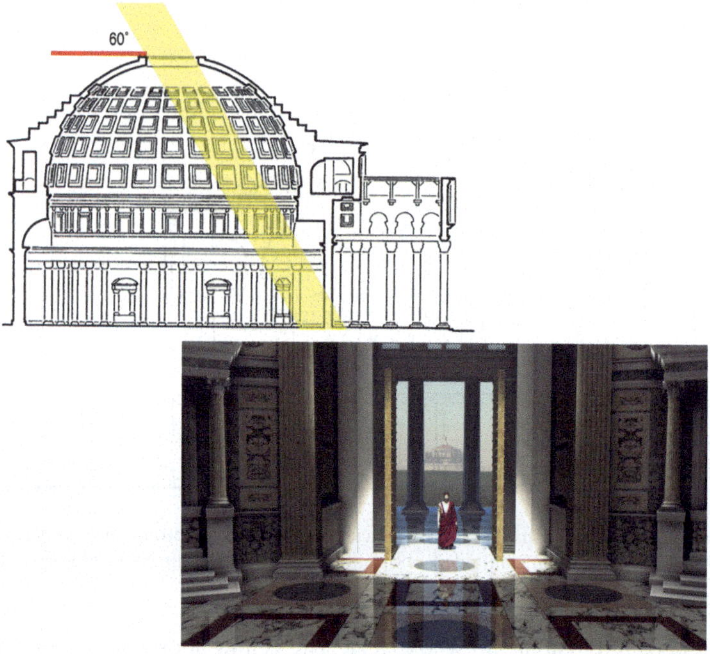

Fig. 3.3 Rome, the Pantheon. On the left, the shaded area shows the sunlight through the oculus at noon on 21 April, when the Sun's altitude is 60° (after Hannah and Magli 2011: 495: Fig. 7). On the right is a virtual reconstruction of entrance of the Emperor into the Pantheon on 21 April (image reproduced by kind permission of John Fillwalk and The History Channel)

The architecture harnesses the noontime sunlight at other moments in the year, most notably at the cusps between one solar/zodiacal month and the next, when the midday sunbeam crosses distinct articulations in the patterned decoration of the floor, the doorway and vertical wall, and the coffers of the domed ceiling, suggesting a coherent organising principle in the design plan that is driven by the Sun. We have seen how it does this at the equinox, harmoniously coordinating mid-points of space and time. That moment in time around 21 March (or correspondingly 23 September, since the sunbeam tracks back this way later in the year as well) lies on the cusp of the zodiacal months of Pisces and Aries (or between Libra and Scorpio). On 21 April, on the cusp between the zodiacal months of Aries and Taurus, the midday Sun shone (and still shines, since the differences in calendar between then and now are negligible) directly on the threshold of the huge doorway entrance to the Pantheon, so that anyone entering the building at that moment was, and is, cast into a spotlight of sunshine (Fig. 3.3). The lower edge of the elliptical beam of light then marks out the inner circumference of the cylindrical wall of the Pantheon. That same point is lit again around 22 May/23 July, but this time by the upper edge of the ellipse of light beaming through the oculus.

This design principle is driven by the Sun itself. But more than that, culturally it may have been more fundamentally driven by calendrical concerns, particularly religious ones associated with Augustus and his family. While the spring equinox seems not to have any significance for Augustus in the calendar, the corresponding date of the autumn equinox more or less coincided with his birthday on 23 September; it was noted in the public calendars (*Fasti*), and was celebrated on that day by the Arval Brethren, the priesthood revived by Augustus, while the public *Fasti* note the equinox on the following day (Degrassi 1963: 512–514; Scheid 1990: 435 (CE 38)). 21 April, at the start of Taurus, was the festival of the Birthday of the city of Rome (Degrassi 1963: 443–445; Hadrian also dedicated his Temple of Venus and Rome on this day). On 23 August, the corresponding, paired date in the zodiacal calendar when the Sun 'enters' Virgo, the Arval Brethren marked the festival of the Volcanalia, in honour of the god Vulcan. 22 May, when the Sun 'entered' Gemini, lies in the period when the Arval Brethren would celebrate their own major festival of Dea Dia (Scheid 1990: 435–439).[4] The corresponding date at the start of Leo, 23 July, lies early in the period between 20 and 30 July, when the *Ludi Victoriae Caesaris* were celebrated (Degrassi 1963: 485–486). These games had been established to commemorate the great multiple triumph of Julius Caesar, Augustus's adoptive father, of 46 BCE. The summer solstice, around 23 June, coincides with a centring of the sunlight's noontime beam on the first circle-in-a-cross feature of the marble floor decoration. This solstice occurred at practically the same time as the festival of Fors Fortuna, the goddess of good fortune, on 24 June (Degrassi 1963: 472–473). Whether any formal relationship was recognised between the festival and the solstice is unknown, but in the grove at La Magliana outside Rome where the Arval Brethren had their main cult site, there was also a temple of Fors Fortuna (Scheid 1990: 150–154). Even though there is no documentary evidence for a link between the two cults of Dea Dia and Fors Fortuna, the coincidence of their cult sites and the revival of the Arval cult by Augustus in this area of Rome seem suggestive of a link.[5] 23 December, the winter solstice when the Sun enters Capricorn, was technically (in astrological terms) the date of the conception of Augustus and we may recall his adoption of the Capricorn as a visual symbol through his reign (Suetonius, *Life of Augustus* 94). It also lay close to the festival of the Birthday of the Sun on 25 December (Degrassi 1963: 545). At this time the second-top row of coffers in the Pantheon's ceiling was illuminated at noon by the Sun; this was the furthest up the ceiling that the Sun shone at midday inside the building. The remaining pairs of zodiacal dates are less illuminating. On 24 October, at the start of the month of Scorpio, the Arval Brethren celebrated a festival for a minor deity,

[4] The dates recorded in the period up to and including Hadrian's time are: 27 and 29 May in CE 38, 19 May in CE 58, 29 May in CE 69, 17, 19 and 20 May in CE 81, 19 and 20 May in CE 87, 19 May in CE 89, 25, 27 and 28 May in CE 90, and 17, 19 and 20 May in CE 105.

[5] Contrast Champeaux 1982: 207–234, who sees Fors Fortuna as a goddess bound to ideas of natural fertility, with Scheid 1990: 715 n. 54, who emphasises instead the goddess's associations with the unexpected military victory of 293 BCE, which gave rise to the establishment of the sanctuary at La Magliana. Later, fourth century CE representations of Fors Fortuna on imperial coins reflect both the watery aspect (in one hand she holds a ship's rudder set on a globe) and the notion of fertility (in her other hand she holds a cornucopia).

Favor (Degrassi 1963: 525), while near the corresponding date of 19 February, the start of Pisces, there was the Feralia on 21 February (Degrassi 1963: 412–414), but this was not celebrated via any public ritual that we know of, instead families made private offerings at their family graves (Varro, *On the Latin Language* 6.13). Finally, 23 November, the start of Sagittarius, presents nothing in the calendars, nor does the corresponding date of 21 January at the start of Aquarius—the commemoration of the birthday of the Divine Hadrian on 24 January is of course a later celebration, after Hadrian's death (Degrassi 1963: 402).

To sum up: in the context of the Pantheon's midday light-show through the year, the equinoctial display looks deliberately contrived at midpoints in space and time. Furthermore, it stands midway between the winter solstice's illumination of the row of coffers in the dome and the summer solstice's illumination of the marble paving in the floor. On both of these occasions the sunlight strikes a clear point of the pattern in the decorative schemes of roof and floor, promoting again the belief that this is all carefully orchestrated in the overall design and decoration of the building. The full explication of the noontime illumination through the other months still eludes us, but it is tempting to see in many of them a reflection of interest in Augustus and his family, especially through the ritual commemorations of the Arval Brethren, the priesthood that he revived and of which he was a member, and which probably met regularly in the Pantheon until CE 63. Whether what we witness in Hadrian's structure is a function of what was visible in Agrippa's in some fashion (it did not have a domed roof, for a start), we do not know, but the festivals noted above certainly go back at least to Augustus's time.

The apparent emphasis in this light-show on the equinoxes and solstices is reminiscent of the cult of Mithras as described by Porphyry and represented in its central icon. This suggests a common underlying view of the cosmos. But can we discern any parallel to the figure of Mithras as the occupier of a position of cosmic balance? For an answer, it may pay to look both backwards and forwards from Hadrian's Pantheon.

Rome: Nero's Domus Aurea

A precedent for the Pantheon's use of sunlight can be found in Nero's Domus Aurea (CE 64–68). The physical remains of the extensive palace are nowadays largely limited to the domestic wing on the Esquiline Hill (Ball 2003; Fabbrini 1995). The orientation of the wing was not a necessary function of the local topography (Voisin 1987). Rather, its strict east-west alignment, which is unique in Imperial structures, must have served some further purpose. This purpose is likely highlighted in the wing's central room, the Octagonal Room.

Here at certain times of the year astronomy defines the room's dimensions: in particular, we may note that the equinoctial midday Sun falls directly on to the north doorway (Fig. 3.4; see Hannah et al. 2016). The lower rim of the ellipse of light cast by the Sun strikes the juncture of the floor and the northern doorway's threshold. So

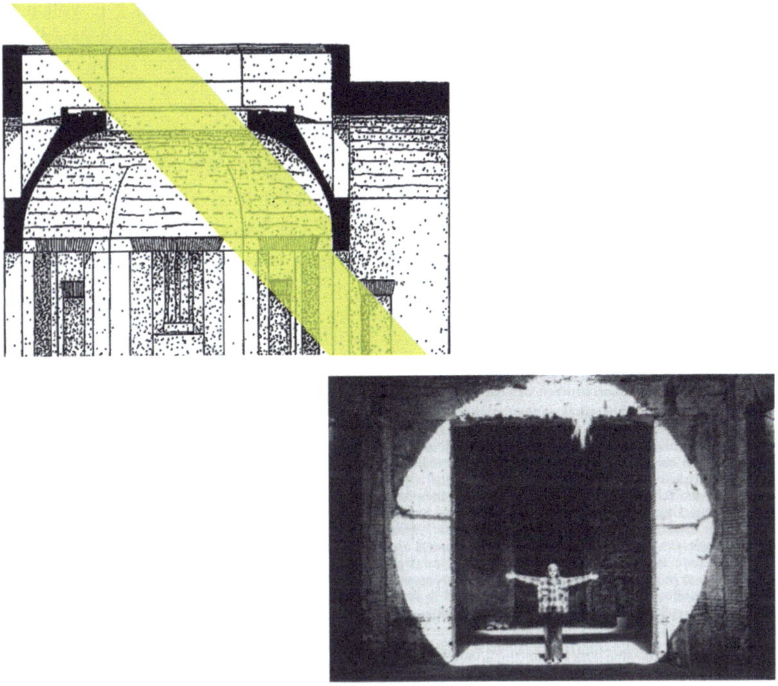

Fig. 3.4 Rome, the Domus Aurea, the Octagonal Room. On the left, the shaded area shows the sunlight through the oculus at noon on the equinoxes, when the Sun's altitude is 48° (after Hannah et al. 2016: 518; Fig. 4); on the right the illuminated area (after Voisin 1987: 516, Fig. 3, reproduced by kind permission of L'École française de Rome)

the Sun measures out the dimensions of the openings in the walls of the Room, in a manner similar to what we saw in the Pantheon.

Nero's association with the Sun is well-attested (Champlin 2003; L'Orange 1947). In the Domus Aurea complex alone there was the colossal statue of the Sun which stood not far from the Octagonal Room (Albertson 2001; Bergmann 1994)—its name is preserved in that of the Colosseum, the amphitheatre which was later built nearby. And in his portraits on coins and in sculpture late in his reign Nero wears the radiate crown that was usually associated with the Sun god (Hiesinger 1975; Smith 2000). We may also note the story of the coronation of Tiridates of Armenia in the Roman forum in CE 66, as told by Dio: Nero sat on the Rostra, facing the Via Sacra, and therefore looking east/southeast, thus also facing the rising Sun—we may imagine Nero's face lit by the Sun—while Tiridates venerated Nero with the words, "I have come to you, my god, to worship you as I do Mithras." (Dio Cassius 63.5.2). For the celebration in the theatre there was an awning overhead which bore an image of Nero driving a chariot, surrounded by golden stars (Dio Cassius 63.6.1–2), an image borrowed from the iconography of the Sun god (Champlin 2003: 276).

The historian Suetonius reports of the Golden House itself:

> [Nero] built a house from the Palatine all the way to the Esquiline, which he called the Passageway House at first, but then, when it was destroyed by fire soon afterwards and rebuilt, the Golden House . . . The main dining hall was circular; it turned round constantly day and night, like the heavens. (Suetonius, *Nero* 31).

The temptation to regard the Octagonal Room as this very dining room has had to be tempered in recent years by the discovery of another structure from the palace over on the Palatine Hill, where a revolving mechanism was a feature (Wilson 2011).

The poet Lucan, writing in Nero's time ca. CE 60, has the Emperor in apotheosis joining the heavens and finding his proper seat midway between the northern celestial sphere and the southern, where he will ensure balance and stability:

> When, with your watch completed, you seek the stars at length, the preferred palace of heaven will welcome you, the sky rejoicing . . . But choose your seat neither in the northern sphere nor where the hot sky of opposed south turns . . . Keep the weight of heaven balanced in the middle of the circle (*orbe . . . medio*); let that part of the clear ether be totally empty and let no clouds obstruct us from Caesar. (Lucan, *The Civil War* 1.45–59).

Here we have a sentiment similar to that given us by Porphyry for the role played by Mithras in his cult's cosmic vision. Both refer to divine figures being situated at celestial points of balance. In Porphyry the god Mithras was placed clearly at the equinoctial region, with the north to his right and the south to his left, overseeing the passage of souls into and out of this world (or out of and into the world of the gods). In Lucan we see the Emperor, apotheosised after death, positioned also somewhere between north and south, although exactly where is left unclear.[6]

In the Octagonal Room, surmounted by its cave-like dome, the equinoctial Sun at noon fully illuminates the northern doorway, which gives entry to a nymphaeum—a space sacred to water nymphs. Let us recall again the associations highlighted by Porphyry between water, nymphs and souls descending into *genesis*:

> Therefore he [Homer] dedicated the gates neither to the east and west nor to the equinoxes, namely Aries and Libra, but to the south and north, and to the gates most southerly to the south and those most northerly to the north, because he dedicated the cave to souls and water nymphs. And for souls these places are fit for generation (*genesis*) and return from generation (*apogenesis*). (Porphyry, *On the Cave of the Nymphs* 24).

Nero's Octagonal Room therefore presents a convincing case in which architectural elements in a major Roman building were symbolically tied to aspects of the Sun's passage through the year. That these associations held metaphysical significance for the Emperor himself, as a divine figure overseeing the cosmos from a point of balance, is then suggested by the contemporary poetic admonition by Lucan to Nero, when he will eventually be apotheosised. So we do indeed find a parallel to the figure of the solar god Mithras: it is the Emperor himself who presents himself in the

[6]Where Lucan imagines the Emperor residing in the Heavens depends on the meaning of the phrase '*orbe . . . medio*': if *orbis* means 'circle' here, then it could be the middle of the equator or the ecliptic or even the Milky Way. If it means 'sphere', then Haskins (1887: 5) has suggested it means ". . . on the surface of the outer sphere formed by the sky at a point vertically above Rome and equidistant from the horizon in every direction . . ."; this is supported most recently by Roche (2009: 142–43); this would be the zenith.

image of the Sun god and who occupies a point of cosmic balance. Can we find these associations elsewhere?

Rome: The Palace of Domitian

Between the Domus Aurea and the Pantheon, are there other buildings that used the Sun in this manner? The great architectural projects in Rome in this period were the Palace of Domitian, and the Forum and Baths of Trajan. Trajan's Forum is dominated by traditional post-and-lintel architecture with no fully domical forms. The Baths certainly used domes, but were a functional, social complex. It is instead with the Palace of Domitian and with the characterisation of that Emperor that we find closer parallels to Nero and his palace. We may also note in passing that Domitian was responsible for one of the rebuildings of the Pantheon.

There are formal architectural similarities with the Domus Aurea, especially in the Octagonal Rooms of the two palaces. There are also parallels in the divine aspirations of the two Emperors, as poets in Domitian's time show, when they aim to capture his godly essence. Statius describes the Olympian dwelling of Domitian on the Palatine, which stuns even Jupiter the Thunderer himself:

> I seem to recline among the stars with Jupiter, and to take immortal wine offered by a Trojan hand . . .
>
> An august building, huge, magnificent, not with a hundred columns but with as many as could support heaven and the gods above, if Atlas were relieved. The neighbouring palace of the Thunderer is stunned, the powers rejoice that you are set in an equal place, nor would you hurry to ascend to the great sky . . .
>
> Far above your sight goes: you would scarcely reach the summit with tired vision, and you would think it the ceiling of the gilded sky. (Statius, *Silvae* 4.2.10–12, 18–22, 30–31).

Martial (*Epigram* 7.56) claims the palace that the architect Rabirius has built has encompassed heaven itself: "You have embraced the stars and the pole in your pious mind, Rabirius . . ." Both poets even call the Emperor a god:

> Behold he [Domitian] is a god, Jupiter commands him to rule the fortunate earth in his place . . .
>
> Hail, leader of men, and parent of gods . . . (Statius, *Silvae* 4.3.128–129, 139).

> Indeed we ask for great things, O gods, but they are what is owed to earth: what prayers are inappropriate for so great a god? (Martial, *Epigram* 4.1).

What is lacking, however, is an appropriate orientation of Domitian's palace, or its domed rooms, to capture the Sun in the same way as the Domus Aurea of Nero or the Pantheon of Hadrian, which frame the Palace chronologically. But perhaps there is more work to be done on this aspect of the building: while a strict cardinal orientation would be ideal, as occurs in the other two structures, nevertheless an effect at a particular time of day or season might have been sought and attainable from the south-westerly angle that these rooms face.[7] We may recall the story of the coronation of Tiridates, when Nero faced the rising Sun and Tiridates honoured him as the equivalent of his god, Mithras. Perhaps something similar was possible in the Palace of Domitian with the westering Sun. Statius (*Silvae* 4.2.38–42) alludes to the Sun-like quality of the Emperor's face:

> But I had no time to gaze at the food or the Moorish wood supported on Indian columns, or the crowd of servants in ranks, so great was the desire to look at him, at him, with calm face, tempering his rays with serene majesty . . . (Statius, *Silvae* 4.2.38–42).

Could some hierophany have occurred with Domitian, for whom far more explicit Sun-god equations were made than even for Nero?

The Orchestration of Time

Much of the foregoing discussion suggests that patrons and architects were interested in directing sunlight for special effects at certain times of the year through the orientation and design of particular buildings. The significant times included the solstices and equinoxes, which are pivotal times in the solar year respectively as the end-points and the mid-points in the cycle of the Sun through the year. These points had metaphysical significance in some religious contexts, as the seat of the Sun god (equinoxes) between gateways for the soul (solstices).

What we can no longer see clearly now is how rituals or other ceremonies were orchestrated so as to create an interaction between the principal actors and the sunlight at these times. We might speculate how the emperor entered the Domus Aurea on special days, spotlit in the sunlight, but this would be only an hypothetical reconstruction with no documentary evidence directly to support it. Similarly, the representation of Hadrian in the doorway of the Pantheon on 21 April in Fig. 3.3 is only an imaginative reconstruction. Can we gain a sense of this interplay elsewhere to help us imagine it and perhaps to help us to interrogate the use of the spaces in these imperial buildings at certain times of the year?[8]

An impression of how light could be used for ceremonial or ritual purposes might come from studying the use of light among the descendants of the early palaces and the Pantheon. These are the Christian churches in western Medieval Europe and in the eastern Byzantine Empire. Here I wish to focus on the Byzantine buildings, but new work is starting to address the rituals that took place in western Medieval churches also

[7]A quick investigation of the Palace on Googe Earth indicates that the orientation is about azimuth 222°, which means that the afternoon Sun would be able to shine on and into the Palace throughout the year.

[8]A ". . . congruency of note, sound, space, and time . . ." has been sought in the Pantheon, but with the acknowledgement that there is no supporting documentary evidence of any rituals that may have taken place in the building: see Wilkins (2004: 82), and Sperling (2004).

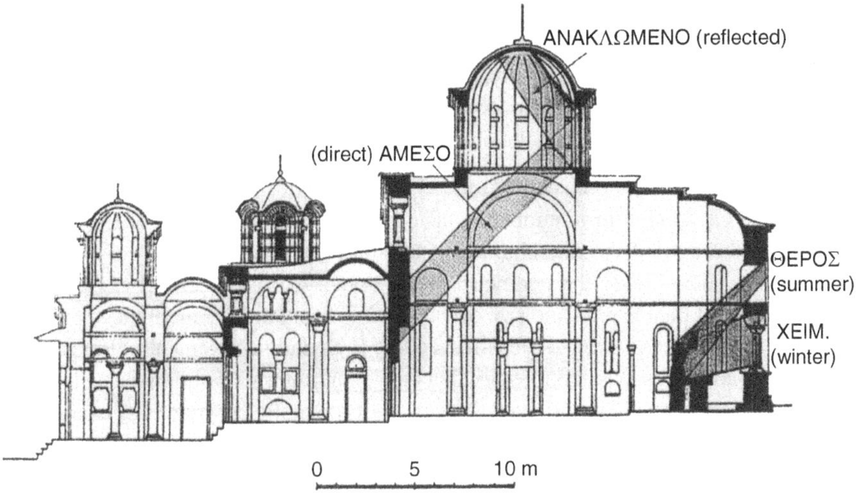

Fig. 3.5 Mount Athos, Monastery of Iveron. The shaded areas show reflected sunlight towards the top of the dome, and direct sunlight on the frescoes in the nave and on the Holy Table (after Potamianos 2000: 46, Fig. E.20, reproduced by kind permission of Professor I. Potamianos)

(e.g. Harper et al. 2016), and the play of light within those buildings as it related to specific feast days in the Christian calendar of the time (e.g. Tardy 2013).

Of the seven types of Byzantine church identified by Patricios, five have in common the dome as an architectural feature. These are the cross-in-square, cruciform, domed basilica, centralised and Athonite. Only the basilica proper and some converted churches omit the feature. If the theory is correct that the imperial palace audience hall was the inspiration for types as divergent as the basilica and the centralised types, then our tour of the Domus Aurea and the Palace of Domitian has introduced us to some of the archetypes available from early in the Roman Empire's architectural lexicon (Patricios 2014: 47–50).

Furthermore, Early Byzantine liturgical ritual (ca. CE 330–610) has been characterised as being 'more classical and more Roman in its sense of Imperial display' and 'conceived as an open action', unlike the later Medieval theatre of 'concealment and revelation' (Mathews 1971: 178). Entrances and processions featured prominently in this liturgical activity, and were regarded by early Byzantine commentators like Maximus the Confessor (seventh century CE) as symbols of the divine will on Earth (Taft 2012: 2–3; for the commentary of Maximus the Confessor, see Berthold 1985: 181–225).

The liturgy performed in these churches also reflects the effects of natural light within the buildings (Fig. 3.5). Following a study of these effects in the churches on Mount Athos, Potamianos recognised three types of observable light-effects:

1. A shaft of light shining on the Holy Sacraments at the time of their dedication to God;

2. The apex of the church dome often appearing radiant with a light that seems to come from the representation of Christ Pantocrator; and
3. The central part of a domed Byzantine church operating as a roofed spherical sundial, in which the cast shadow that indicates the time of day is replaced by a spot of light from the top of the sundial.

Ensuring the continuous realisation of the light-effects in Byzantine churches required the calculation of a number of factors, from the computation of the position of the Sun throughout the day and year particularly in relation to holy days, to the orientation of the church, the geometry of the dome, the placement of the altar, the determination of the location and size of the windows, and the timing of the significant moments in the liturgy. These light-effects occur both daily and annually. Imagery, in the form of icons, frescoes and mosaics that have major liturgical significance, tends to be found in areas that are highlighted daily by spots of sunlight, while images of lesser importance may be highlighted just once a year (Potamianos 1996: 20–21).

In a study that focussed solely on the church of Hagia Sophia in Istanbul, Jabi and Potamianos argued persuasively for a conscious integration by the architects of geometry, light and cosmology in the building. The apse, for example, was so designed as to allow a shaft of light to fall on the altar through one of the apse windows at the Byzantine third hour, during which the Holy Spirit, represented by the beam of light, was imagined as descending to sanctify the offerings presented to God by the priest in front of the altar. Also of interest here is that although it can be shown that the church is oriented on its longitudinal axis towards sunrise at the winter solstice,[9] nonetheless Jabi and Potamianos show that in terms of the ritual within the church, the orientation also makes sense if understood as representing the Sun's azimuth at the third hour of different days, notably 25 December (Julian), which is Christmas Day, *and* the equinox, with the beam of sunlight being filtered through different windows in the apse on each occasion (Jabi and Potamianos 2007: 307, 309). This suggests that orientations were multivalent, being significant for different days and particular hours of the day, and not only for sunrise along the horizon.

Some of this planning we have seen applied also to the sacred and secular buildings of imperial Rome: orientation of the building, geometry of the domed rooms, scale of the oculus (in place of the later windows). There is no altar, and we do not have the equivalent of the liturgy, but there may be aspects of the later Christian ceremonial that could help us to imagine the entrance of the Emperor, in a kind of hierophany—if not in fact a theophany, given the divine nature accorded some of the Emperors.

The second of Potamianos's light-effects, the apparently constant radiation of light from the dome and particularly from the dominant image of the Pantocrator in the dome, is achieved via a design feature in the dome's structure that seems not to have been utilised in the imperial Roman buildings that we have been studying.

[9]Albeit at the slightly inaccurate latitude that was understood to apply to the city at the time: Schibille (2009).

While some of this effect may be achieved through careful shaping of the curvature of the dome and the use of reflected light from the dome's windows, sometimes a special contrivance in the form of a reflecting mirror was utilised to increase the effect and make it constant through the day to allow for the changing position and altitude of the Sun (Jabi and Potamianos 2007: 310–316; Potamianos 1996: 120–131). The only potential instance of the use of reflected light in this manner (but certainly not of a mirror) that I am aware of would be if the dome of the Domus Aurea was in fact topped by a lantern, which was punctuated by small windows, as was once suggested (Hemsoll 1989). But such an addition now seems improbable.[10]

The first and the third effects listed by Potamianos, however, are relevant to our present purposes. We have seen how the Pantheon may be viewed as the equivalent of a roofed sundial, at least for the period of noon through the year. The equivalent of the shaft of light striking the bread and wine of the Eucharist at the particular moment of dedication would be not a daily event but a biannual one in the case of the noontime Sun striking the northern doorway of the Octagonal Room in the Domus Aurea at the time of the equinoxes, or an annual one for the beam of sunlight in the Pantheon striking the doorway at noon on the occasion of the Birthday of Rome. We might posit a significant figure, such as the Emperor, in the doorways at that moment. The light-show at the equinoxes in the Pantheon must have been significant, since the beam of sunlight strikes a significant spatial point (halfway up the building, at the juncture of the artificial hemispherical dome and the cylindrical wall) at a significant temporal point (halfway through the day, halfway through the solar cycle between solstices). But what was the significance of this coincidence culturally?

In the case of the Domus Aurea, we are stymied by a lack of documentary evidence for our suspicion that the light-effect at noon at the equinoxes was meaningful, despite a carefully designed coincidence of lighting and architectural design. There is no formal link in the festival calendar between the equinoxes and events in the life of Nero, such as his *dies natalis* (which was celebrated on 15 December) or his accession to Emperor (13 October). At most, the spring equinox lay close to the time of Nero's supposed conception (~15 March), and it is not impossible that Nero considered this significant, given the precedence of Augustus' practice. Augustus, who was born near the autumn equinox and therefore conceived around the time of the winter solstice when the Sun was 'in' Capricorn, chose to promote the sign of Capricorn in his public imagery. With Nero, as we have seen, the poet Lucan had the Emperor in apotheosis finding his proper seat on a celestial mid-point, where he would ensure balance and stability.

We have also seen such a point of balance was adopted by the followers of Mithras: the god himself was placed in the equinoctial region, where he could oversee the descent of souls into *genesis* into this world, and their ascent in *apogenesis* into the immortal world. That such an interpretation may be attributable

[10]Ball (2003: 215) makes no mention of this possibility in his description of the structure of the dome. Nonetheless he does note the extensive use of reflected light in the Octagon Suite.

also to Nero is not far-fetched, as is demonstrated by the solar aspects that the Emperor himself adopted in his imagery in coinage and sculpture, and even more tellingly by the story related by Cassius Dio of the coronation of the Persian Tiridates in CE 66, when Nero was equated with Mithras (see further Hannah et al. 2016: 520–522).

The divine status that we see here accorded the Emperor before death, while common enough in the Hellenistic East, was dangerous in the Roman West, and yet we see it further enlarged with Domitian. Poets of his time seem to have vied with each other in rendering praise to the Emperor in increasingly divine terms, equating Domitian with Jupiter himself, and his palace with Olympus. What we cannot yet discern, however, is to what degree, if at all, Domitian's palace also provided a light-show in terms similar to that displayed in the Domus Aurea, especially at the equinoxes.

Domitian's palace remained the imperial seat in Rome for later Emperors like Hadrian, but he spent relatively little time there as opposed to travelling to the east of the Empire or residing in his extensive Villa at Tivoli. There we now see, with increasing certainty, the use of solar alignments in parts of the Villa (Castellani 2006; Frischer et al. 2016). To what extent this is transferrable to the Pantheon, as seat of the Emperor in certain public duties, is again harder to be sure of, but there do appear to be grounds for seeing the building as a site for an imperial hierophany at certain times of the year that are significant astronomically and, to some extent, culturally.

Acknowledgement I would like to thank Professor Giulio Magli for the invitation and material support that enabled me to participate in this conference. This research was also supported by a Marsden Fund grant from the Royal Society of New Zealand (10-UOO-064).

References

Albertson, F. C. (2001). Zenodorus's 'Colossus of Nero'. *Memoirs of the American Academy in Rome, 46*, 95–118.

Ball, L. F. (2003). *The Domus Aurea and the Roman architectural revolution*. Cambridge: Cambridge University Press.

Beard, M. (1985). Writing and ritual: A study of diversity and expansion in the Arval Acta. *Papers of the British School at Rome, 53*, 114–162.

Beaujeu, J. (1955). *La Religion romaine à l'Apogée de l'Empire. I. La Politique religieuse des Antonins*. Paris: Les Belles Lettres.

Beck, R. (2006). *The Religion of the Mithras Cult in the Roman Empire*. Oxford: Oxford University Press.

Bergmann, M. (1994). *Der Koloss Neros: Die Domus Aurea und der Mentalitätswandel im Rom der früher Kaiserzeit*. Mainz am Rhein: P. von Zabern (Trierer Winckelmannsprogramme, 13).

Berthold, G. C. (1985). *Maximus confessor: Selected writings*. New York: Paulist Press.

Castellani, V. (2006). Tivoli: Villa Adriana, Rocca Bruna e l'astronomia. *Rivista Italiana di Archeoastronomia, 4*, 9–18.

Champeaux, J. (1982). *Fortuna. Le Culte de la Fortune à Rome et dans le Monde romain. I – Fortuna dans la Religion archaïque*. Rome: École française de Rome.

Champlin, E. (2003). *Nero*. Cambridge (Mass.): Belknap Press of Harvard University Press.

Coarelli, F. (1983). Il Pantheon, l'Apoteosi di Augusto e l'Apoteosi di Romolo. In K. de Fine Licht (Ed.), *Città e Architettura nella Roma Imperiale. Analecta Romana Instituti Danici: Supplementum X* (pp. 41–46). Odense: Odense University Press.
Degrassi, A. (Ed.). (1963). *Inscriptiones Italiae XIII.2*. Rome: Istituto Poligrafico dello Stato.
Fabbrini, L. (1995). Domus Aurea: Il Palazzo sull'Esquilino. In E. M. Steinby (Ed.), *Lexicon Topographicum Urbis Romae* (Vol. 2, pp. 56–63). Rome: Edizioni Quasar.
Frischer, B., Zotti, G., Mari, Z., & Capriotti Vittozzi, G. (2016). Archaeoastronomical experiments supported by virtual simulation environments: Celestial alignments in the Antinoeion at Hadrian's Villa (Tivoli, Italy). *Digital Applications in Archaeology and Cultural Heritage, 3*, 55–79.
Hannah, R. (1996). The image of Cautes and Cautopates in the Mithraic tauroctony icon. In M. Dillon (Ed.), *Religion in the ancient world: New themes and approaches* (pp. 177–192). Amsterdam: Hakkert.
Hannah, R. (2009a). *Time in antiquity*. London: Routledge.
Hannah, R. (2009b). The Pantheon as a timekeeper. *British Sundial Society Bulletin, 21*(4), 2–5.
Hannah, R. (2019). From here to the hereafter: Genesis and apogenesis in ancient philosophy and architecture. In I. Ramelli & S. Slaveva-Griffin (Eds.), *Lovers of the soul, lovers of the body: Perspectives on soul and body and soteriology in late antique Platonism*. Washington: Harvard University (Centre for Hellenic Studies).
Hannah, R., & Magli, G. (2011). The role of the Sun in the Pantheon design and meaning. *Numen, 58*(4), 486–513.
Hannah, R., Magli, G., & Palmieri, A. (2016). Nero's 'solar' kingship and the architecture of the Domus Aurea. *Numen, 63*, 511–524.
Harper, S., Barnwell, P. S., & Williamson, M. (Eds.). (2016). *Late Medieval liturgies enacted: The experience of worship in Cathedral and Parish Church*. Farnham: Ashgate.
Haskins, C. E. (1887). *M. Annaei Lucani Pharsalia*. London: George Bell and Sons.
Hemsoll, D. (1989). Reconstructing the octagonal dining room of Nero's Golden House. *Architectural History, 32*, 1–17.
Hiesinger, U. W. (1975). The portraits of Nero. *American Journal of Archaeology, 79*, 113–124.
Iara, K. (2015). Moving in and moving out: Ritual movements between Rome and its *Suburbium*. In I. Östenberg, S. Malmberg, & J. Bjørnebye (Eds.), *The moving city: Processions, passages and promenades in ancient Rome* (pp. 125–132). London: Bloomsbury.
Jabi, W., & Potamianos, I. (2007). Geometry, light, and cosmology in the Church of Hagia Sophia. *International Journal of Architectural Computing, 5*(2), 303–319.
L'Orange, H. P. (1947). *Apotheosis in ancient portraiture*. Oslo: H. Aschehoug.
Lange, C. H. (2015). Augustus' triumphal and triumph-like returns. In I. Östenberg, S. Malmberg, & J. Bjørnebye (Eds.), *The moving city: Processions, passages and promenades in ancient Rome* (pp. 133–143). London: Bloomsbury.
Mathews, T. F. (1971). *The early churches of Constantinople: Architecture and liturgy*. University Park: Pennsylvania State University Press.
Patricios, N. N. (2014). *The sacred architecture of Byzantium: Art, liturgy and symbolism in early Christian churches*. New York: I.B. Tauris.
Potamianos, I. (1996). *Light into architecture: Evocative aspects of natural light as related to liturgy in Byzantine churches*. Ph.D. thesis. University of Michigan.
Potamianos, I. (2000). *To Phos ste Vyzantine Ekklesia*. Thessalonike: University Studio Press.
Roche, P. (2009). *Lucan, De Bello Civili Book I*. Oxford: Oxford University Press.
Scheid, J. (1990). *Romulus et ses Frères: Le Collège des Frères arvales, Modèle du Culte public dans la Rome des Empereurs*. Rome: École Française de Rome.
Schibille, N. (2009). Astronomical and optical principles in the architecture of Hagia Sophia in Constantinople. *Science in Context, 22*(1), 27–46.
Smith, R. R. R. (2000). Nero and the Sun-god: Divine accessories and political symbols in Roman imperial images. Review of Marianne Bergmann, *Die Strahlen der Herrscher: Theomorphes Herrscherbild und politische Symbolik im Hellenismus und in der römischen Kaiserzeit. Journal of Roman Archaeology, 13*, 532–542.

Sperling, G. (2004). Early orchestration: The Pantheon as a resonance element. *Nexus Network Journal, 6*, 65–69.
Stahl, W. H. (1952). *Macrobius, commentary on the Dream of Scipio*. New York: Columbia University Press.
Taft, R. (2012). Liturgy. In R. Cormack, J. Haldon, & E. Jeffreys (Eds.), *The Oxford Handbook of Byzantine studies*. https://doi.org/10.1093/oxfordhb/9780199252466.013.0055
Tardy, D. (2013). *Toute la Lumière sur l'Église romane de Saint-Nectaire: Tome 1, Les Découvertes (2009–2012)*. Lyon: Éditions BZT et Cie.
Voisin, J.-L. (1987). *Exoriente Sole* (Suétone, *Ner.* 6): D'Alexandrie à la Domus Aurea. In *L'Urbs: Espace urbain et Histoire (1er Siècle av. J.-C. — IIIe Siècle ap. J.-C.): Actes du Colloque international, Rome (8–12 Mai 1985)* (pp. 509–541). Rome: École française de Rome.
Wilkins, P. (2004). The Pantheon as a globe-shaped conception. *Nexus Network Journal, 6*, 78–84.
Wilson, R. J. A. (2011). Neue Forschungen an der Domus Aurea. *Antike Welt, 5*(11), 6.

Chapter 4
The So Called Neo-Pythagorean Basilica of Porta Maggiore in Rome: The Most Mysterious Roman Monument

V. F. Polcaro, S. Sclavi, S. Gaudenzi, L. Labianca, and M. Ranieri

Abstract The underground Basilica of Porta Maggiore, accidentally discovered in 1917, is not mentioned in any classical source. All the authors who have studied the monument agree on dating it to the first half of the first century AD, independently from the assumptions made on its intended use. The Basilica is in a good state of conservation, and preserves most of its original decoration. After discussing briefly the main features of the monument, we shall examine it from an archaeoastronomical point of view, noting a very interesting feature: that at summer solstice sunset an optical path illuminates one of the scenes decorating the Basilica vault. The

V. F. Polcaro (✉)
INAF, Istituto Nazionale di Astrofisica, Rome, Italy

IAPS, Istituto di Astrofisica e Planetologia Spaziali, Rome, Italy

Astronomy and Cultural Heritage ACHe Center, Ferrara, Italy

Università degli Studi di Ferrara, Ferrara, Italy

Associazione Culturale C.E.S.A.R., Center of Excellence Studies and Archaeological Research, Rome, Italy
e-mail: info@cesar.roma.it; vitofrancesco.polcaro@iaps.inaf.it

S. Sclavi
Università di Roma "La Sapienza", Rome, Italy
e-mail: silvia.sclavi@uniroma1.it

S. Gaudenzi
INAF, Istituto Nazionale di Astrofisica, Rome, Italy

IAPS, Istituto di Astrofisica e Planetologia Spaziali, Rome, Italy

Università di Roma "La Sapienza", Rome, Italy
e-mail: Silvia.Gaudenzi@uniroma1.it; silvia.gaudenzi@iaps.inaf.it; Silvia.Gaudenzi@roma1.infn.it

L. Labianca
Soprintendenza Archeologica di Roma, Rome, Italy
e-mail: lucilla.labianca@fastwebnet.it

M. Ranieri
INAF, Istituto Nazionale di Astrofisica, Rome, Italy

IAPS, Istituto di Astrofisica e Planetologia Spaziali, Rome, Italy

G. Magli et al. (eds.), *Archaeoastronomy in the Roman World*, Historical & Cultural Astronomy, https://doi.org/10.1007/978-3-319-97007-3_4

centrality of this hierophany will hopefully help to disentangle the many mysteries of this monument.

The Monument

The so called Neo-Pythagorean Basilica of Porta Maggiore in Rome was discovered on 23 April 1917 after a collapse of the ground under the Rome-Cassino railway line. The monument lay at more than 10 m below the level of the Via Prenestina, just outside Porta Maggiore (Gatti and Fornari 1918); since it was not mentioned in any of the known sources, nobody suspected its existence before the discovery.

The Basilica was designed to be underground since its creation. In fact, it was never 'built', but was dug 'in negative' (Peschi 2007). First the pillars and the outer walls were excavated, and were later filled with a concrete casting of lime and pozzolan with fragments of flint. The construction continued with a coarse ribbing of the arches and vaults with lumber, and by shaping the excavated earth. Finally, the construction ended with the direct casting of the vaults. Once the structure of the basilica was built in negative, the earth still found inside was removed.

However, at the time of discovery, the Basilica was found nearly completely filled with earth. This filling appears intentional and not the result of a collapse of the surrounding soil: in ancient times, earth had been thrown into the monument, erasing all external traces of it.

The complex of the Basilica (Aurigemma 1974; see Fig. 4.1) consists of an access corridor (of which today only the last part remains) that leads to a small passageway or vestibule, from which one enters into the main hall of the Basilica. In the vestibule there is a skylight, likely designed both for air exchange and for other purposes, which we shall discuss later. The hall of the Basilica is divided into three naves, covered with barrel vaults. The central nave is wider than the side ones and has an apse at the bottom.

In Roman times the area where the Basilica is located was a suburban one, called 'ad Spem Veterem', given the existence of an archaic temple dedicated to Spes (i.e. Hope), called Vetus to avoid confusion with the temple built later, dedicated to the same deity, but located in the Forum Olitorium.

The Basilica Decoration

The Basilica walls and vaults are decorated with polychrome frescoes and white stuccos. The two main scenes seem to be the one in the apse, interpreted as Sappho in the act of throwing herself from the Lefkada cliff, and the one at the center of the vault of the central nave, interpreted as the 'Rape of Ganymede' (Fig. 4.2).

On the rest of the surfaces, molded frames delimit geometrical boxes with figurative representations from the repertoire of classical mythology (Apollo and

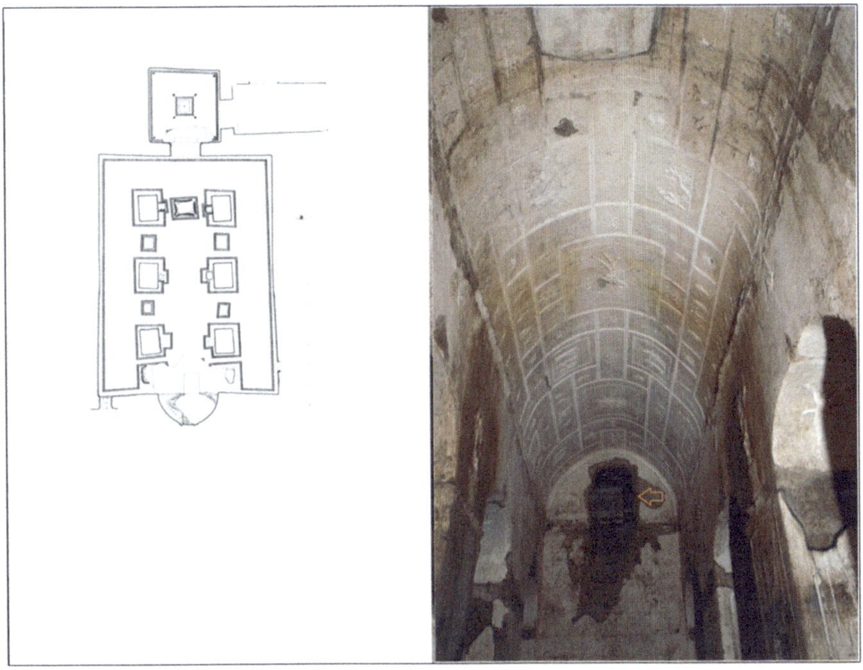

Fig. 4.1 On the left is a plan of the Basilica (from Aurigemma 1974). On the right is a photograph of the central nave of the Basilica, seen from the apse. The window on the wall between the nave and the vestibule is shown by the orange arrow (courtesy: Rome Archaeological Superintendence)

Fig. 4.2 Left: The 'Rape of Ganymede'. Right: The 'Suicide of Sappho' (courtesy: Rome Archaeological Superintendence)

Fig. 4.3 The pigmy returning from a hunting expedition. Notice the East African tucul and the coconut palm on the right (courtesy: Archaeological Superintendence of Rome)

Marsyas, Agave with the head of Pentheus, Apollo and a muse, a satyr abducting a nymph) and from mystical rituals. Female figures making an offer are frequent as well as images of vases, candlesticks, musical instruments, laid tables and other objects, all of which clearly take on a symbolic meaning. Large panels with stylized landscape frescoes are also present.

Scenes apparently taken from everyday life, such as the representation of a school, are less understood in this context, and the scene of a pygmy who returns from a hunting expedition (Fig. 4.3) is even less easy to interpret. With regard to this scene, many curious features are worthy of notice. The pygmy is represented in the usual form of a monstrous dwarf, common in many of the so-called 'Nilotic scenes' frescoes of the Early Imperial Period. However, his hut is the exact representation of the ones still in use today, and typical of the Equatorial Eastern African coast (the tucul of Somaliland and Kenya).

Furthermore, the palm tree in the background is not a date palm, common in Northern Africa, but a coconut palm, again typical of Equatorial Eastern Africa, and not present in the Mediterranean area because of climatic reasons (Gunn et al. 2011). It would seem, therefore, that the painter, although he had never seen an actual pygmy,[1] was somehow aware of the actual landscape of the east coast of Central Africa, even though no Roman expedition reaching as south as Somaliland or Kenya is known.

[1]In fact, this ethnic group is not present in East Africa, but only in Central Africa.

The Hypotheses on the Use of the Basilica

The discovery of the monument aroused great interest, and Lanciani (1918) defined it as "... unparalleled".

The interpretations proposed by scholars regarding the intended use of the Basilica have been many, but basically can be divided between two hypotheses: that it was either a funerary monument or the place for a mystery cult. However, all the authors who have studied the monument, irrespective of the assumptions made on its intended use, agree on dating the monument to the first half of the first century AD, due to clearly identifiable decorative subjects and styles.

Carcopino (1927) suggested that the complex could be attributed to the properties of Statilii and in particular to the possessions of T. Statilius Taurus, consul in AD 44, who committed suicide in AD 53 following the accusation of magical practices, as reported by a brief passage from the Annales of Tacitus (XII, 59). Carcopino (following a suggestion by Gatti and Fornari 1918) based his interpretation on a passage from Frontinus, who, referring to the temple of Spes Vetus, identifies it as "... in confinio Horthorum Torquatianorum et ... norum". Carcopino integrates the gap as *Taurianorum*. In fact, the underground tomb of the Statilii family freedmen is sited just two hundred meters from Porta Maggiore, currently under the building of the main FIAT dealership in Rome.

Carcopino, developing the interpretation of the monument given by Cumont (1918), identified the basilica as a place of the neo-Pythagorean cult. According to these scholars, the architectural layout follows precisely the dictates of this doctrine in the orientation and plan, which includes a long corridor coming from the East. In this interpretation, the decoration prefers the white color in the hall since it is the symbol of the purity of the soul, through which the contact with the divine becomes possible. Carcopino also suggested that the stucco in the center of the main vault represents the 'Rape of Ganymede', interpreted as the representation of an ideal world, where life turns toward the celestial sphere. He also interpreted the subject of greater proportions chosen for the apse as the 'Suicide of Sappho', alluding to the salvation reserved for followers after death.

The hypothesis on the use of the Basilica as the site of a neo-Pythagorean cult could be supported by the fact that the plan of the Basilica is completely based on Pythagorean triads (i.e. all measures, in Roman cubits, perfectly satisfy the Pythagoras theorem; Labianca et al. 2009). However, this was the usual technique to obtain right angles in antiquity (Ranieri 1997). Also, no explicit Pythagorean symbols, such as the 5-points stars included in the pentagon, or scenes attributable to Pythagoras' life or legends are represented in the Basilica decoration.

Indeed, Carcopino's interpretation of both the main figures in the Basilica decoration gives rise to many doubts. If we look at the scene represented in the apse, we wonder whether Sappho would actually be happy to be greeted in the afterlife by two strong men, one ready to receive her with a towel, the other playing the lyre (Fig. 4.4). Furthermore, Zeus abducting Ganymede is never represented as a winged genius, but more like an eagle.

Fig. 4.4 The stucco in the apse of the Basilica (courtesy: Rome Archaeological Superintendence)

Other eminent scholars, such as Bendinelli (1927) and Sauron (1980), identify the monument as a tomb. In particular, Sauron (1982) suggested that the monument is the tomb of another Statilius Taurus, consul in AD 11, and one of the main collaborators of Augustus, of the same 'gens' but of a generation older than that of Statilius Taurus who died because of Claudius' enmity.[2]

In favour of the hypothesis involving Statilius Taurus, who committed suicide in AD 53, is the fact that, at the end of a short period of use (as shown by the state of conservation of the delicate decoration; Labianca et al. 2010), the monument was deliberately abandoned. It was deprived of all internal furniture, then it was filled with earth, and any external structure which could testify its presence was demolished, including those that necessarily covered, or at least marked, the skylight. Such a deed (although not documented by any historical text) is an obvious '*damnatio memoriae*', not justifiable for the tomb of a famous person, but understandable for a place where people have been practicing condemnable magic cults.

An important detail is given by the presence of previous frescoes, visible in several points under the present decorations on the walls, which suggest two phases of use of the Basilica. The change in use must have taken place after a short time lapse, between the era of Augustus and that of Claudius.

[2]In fact, according to Tacitus, because of Agrippina's enmity, who sued him for 'magical practices', to take possession of his estates.

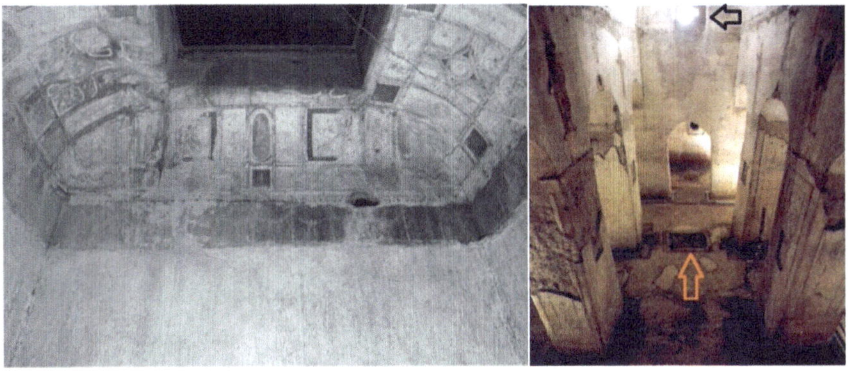

Fig. 4.5 Left: the skylight in the vestibule. Right: The wall between the nave and the vestibule. Notice the hopper window (black arrow) and the imprint of the altar (orange arrow) on the floor between the two first pillars (courtesy: Rome Archaeological Superintendence)

The Archaeoastronomical Analysis

The Basilica hall was totally devoid of openings. Natural lighting and ventilation were ensured through an opening made in the gavetto vault in the vestibule (2.42 × 1.56 m), an opening so shaped as to reproduce in scale the shape of the Basilica. From this opening the external light entered, shedding a diffuse light in the square atrium in front of the hall. From there, the light could faintly illuminate the Basilica through an absided hopper window (measuring approximately 1.10 × 0.80 m), built in the wall separating the atrium and the main hall (Fig. 4.5).

An important feature in the Basilica is the altar that was once located at the entrance of the nave. It was taken away when the Basilica was obliterated, but its former presence is witnessed by the hole it left and by a black tiled border on the mosaic floor. Now, the center of this altar, the center of the opening in the wall (between the atrium and the Basilica) and the center of the skylight in the vestibule lie exactly on the same vertical plane. Thus, the beam of light coming from the opening fell exactly on the altar. The height of the altar was probably 90 cm, equal to that of the small pillars placed against the main nave pillars (of which only the track on the plaster is left). The analysis of the geometry of the Basilica (Labianca et al. 2009) determined that the azimuth of the Basilica axis is 273.0° ± 1.5°. The sky area visible through the skylight from the inside of the Basilica was delimited by the azimuths of 269.6° ± 0.9° and 283.0° ± 0.7° and a height of 34.2° ± 0.2° and 69.7° ± 0.1°. Therefore, the Sun passed within the field of view of the window for about 30 min, just before sunset, during a few days around the summer solstice. The Moon was also visible for a slightly longer period around the Major Northern Lunistice.

Our study of the Basilica showed another interesting detail: because of the particular shape of the window between the vestibule and the hall, this opening is viewed from the altar like a circle. The light beam then outlines a perfectly circular

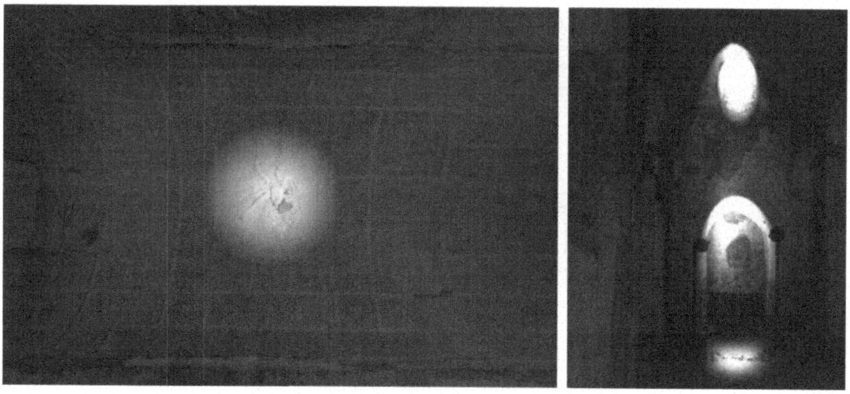

Fig. 4.6 The illumination experiment of the Basilica. Left panel: the circular spot reflected on the 'Rape of Ganymede' stucco. Right panel: the spot on the imprint of the altar (Sciortino and Labianca 2010)

spot on the altar. This effect has been verified by illuminating the skylight with a 500 W headlight (Sciortino and Labianca 2010). In addition, assuming that a reflective surface (e.g. a polished metal basin filled with water) was placed on the altar, this spot would be projected on the precise center of the nave, exactly lighting up the stucco called the 'Abduction of Ganymede' (as demonstrated by the experiment of illumination, see Fig. 4.6). The same effect was surely present when the Sun or the Moon passed through the skylight.

Therefore the skylight in the atrium and the window between it and the hall had not only a practical purpose, but also a ritual one, as a signal of a particular time at which most probably special celebrations were planned.

In order to check whether the visible sky window from inside the Basilica was also suitable for the observation of other celestial bodies, further calculations were performed by Silvia Sclavi as part of her doctoral thesis on "Science for Cultural Heritage" at Rome's 'La Sapienza' University (Sclavi 2016). Specifically, she examined whether it was possible to observe planets or some constellations that may be depicted in the decorations of the Basilica, for example Taurus and Gemini. Using the positional astronomy software Solex V11.0[3], she checked which Solar System bodies visible to the naked eye may have passed in front of the window in the years in which the monument was most probably in use, i.e. between AD 10 and 53.

She found that, in addition to the Sun, only Mercury and Venus were visible more or less all the years of interest, although for short periods only. All others planets and the Moon were only occasionally visible. It is therefore not likely that they were subjects of observation.

With regard to the stars, we must take into account that the field of view is rather narrow, so that very few small constellations were completely visible and were

[3]http://chemistry.unina.it/~alvitagl/solex/

therefore easily recognizable. In addition, only a few bright stars (of visual magnitude brighter than 3) were visible through the window. It is therefore unlikely that stellar observations were performed from the interior of the Basilica.

Conclusions

An archaeoastronomical study of the Basilica has pointed out a feature—the complex light system illuminating the Ganymede scene—that is important in itself as a demonstration of the astronomical notions of the builders. For what concerns the understanding of the purpose of the Basilica, the illumination of the mentioned scene stresses its central role in the cult performed in this structure. The main purpose of the window was to give rise to a hierophany at sunset, the effect being stronger and unmistakable at summer solstices, when the Sun passed directly across the window. Possibly, some effect was visible throughout the year due to the diffused light in the sky.

Of course, there are still many open questions, the answers to which can only be given by archaeology and history: Was the Basilica at first the tomb of Statilius Taurus senior, then becoming the site of the 'magic cults' of the unfortunate Statilius Taurus junior? Were these 'cults' really neo-Pythagoreans? How did the decorator of the Basilica know about coconut palms and East African huts, and what did pygmies have to do with mystery cults? Why did the skylight repeat the plan of the Basilica? Was Sappho really the woman who seems to dive off the cliff, certainly a central character of the cult? Was Ganymede really the character in the arms of the winged genius?

The last question is perhaps the most important one, because, as mentioned earlier, our archaeoastronomical study has shown that this scene was the most important of the cult, since the whole design of the Basilica was directed towards making possible the astonishing hierophany at summer solstice sunset.

References

Aurigemma, S. (1974). *La Basilica Sotterranea Neopitagorica di Porta Maggiore in Roma*. Roma: Istituto Poligrafico dello Stato.

Bendinelli, G. (1927). Il monumento sotterraneo di Porta Maggiore in Roma. *Monumenti Antichi, Pubblicati per cura della Reale Accademia dei Lincei, XXXI*, 601–859.

Carcopino, J. (1927). *La Basilique Pythagoricienne de la Porte Majeure*. Paris: L'Artisan du Livre.

Cumont, F. (1918). La Basilique Souterraine de la Porte Majeure. *Revue Archeologique, 8*, 52–75.

Gatti, E., & Fornari, F. (1918). Brevi Notizie Relative alla Scoperta di un Monumento Sotterraneo Presso Porta Maggiore. *Notizie degli Scavi*, 30–52.

Gunn, B. F., Baudouin, L., & Olsen, K. M. (2011). Independent origins of cultivated coconut (*Cocos nucifera* L.) in the Old World tropics. *PLoS One, 6*(6), e21143. https://doi.org/10.1371/journal.pone.0021143.

Labianca, L., Sciortino, I., Gaudenzi, S., Patané, A., Polcaro, V. F., & Ranieri, M. (2009). An archaeoastronomical study of the 'Neo-pythagorean Basilica' at Porta Maggiore in Rome. *Archaeologia Baltica, 10*, 195–199.

Labianca, L., Massa, S., Gaudenzi, S., Ranieri, M., Polcaro, V. F., & Sclavi, S. (2010). The hypogeal 'Neo-pythagorean Basilica' at Porta Maggiore in Rome. In T. S. Akashesh & M. S. Quintero (Eds.), *Proceedings of the seventh international conference on science and technology in archaeology, and conservation, December 7–11, 2010, Amman, Petra–Jordan* (pp. 298–308). Amman: CulTech for Heritage and Conservation.

Lanciani, R. (1918). Il santuario sotterraneo recentemente scoperto ad Spem Veterem. *Bullettino della Commissione Archeologica Comunale di Roma, XLVI*, 69–84.

Peschi, D. A. (2007). *La Basilica Sotterranea di Porta Maggiore a Roma: "Il cielo celato"*. M.A. thesis. Faculty of Humanities, "La Sapienza" University of Rome, Italy.

Ranieri, M. (1997). Triads of integers: How space was squared in ancient times. *Rivista di Topografia Antica. The Journal of Ancient Topography, VII*, 209–244.

Sauron, G. (1980). Templa Serena. *Mélanges de l'École française de Rome, 92*, 277–301.

Sauron, G. (1982). Discours simbolique et formes décoratives à Rome. *Mélanges de l'École française de Rome, 94*, 699–713.

Sciortino, I., & Labianca, L. (2010). *Osservazioni sulla Basilica sotterranea di Porta Maggiore in Roma*. Atti del VII Convegno SIA "Il Cielo e l'Uomo: problemi e metodi di Astronomia Culturale", Roma, Museo Nazionale Romano, Terme di Diocleziano, 28–29 Settembre 2007, Editore Quasar, Roma.

Sclavi, S. (2016). *L'importanza di un approccio multidisciplinare coordinato nelle indagini su monumenti dell'antichita'*. Ph.D. thesis on Science for Cultural Heritage, 'La Sapienza' University of Rome, Italy.

Chapter 5
The Mausoleum of Santa Costanza in Rome: A Survey of the Light Phenomena Through the Centuries

Flavio Carnevale and Marzia Monaco

Abstract The Mausoleum of Santa Costanza is a perfect example of early Christian architecture during the late Roman Empire, when the Christian cult had just been liberalized. Located in Rome on the Via Nomentana, this circular building belongs to a monumental complex linked to the worship of the tomb of the martyr Saint Agnes, over which two basilicas were built. The oldest of the two, dated to the fourth century, was a horseshoe-shaped basilica that eventually fell into ruins. Being too large to be refurbished, in the seventh century it was replaced by the current much smaller Basilica of Sant'Agnese fuori le Mura.

According to tradition, the Mausoleum was built by Constantina for herself; the daughter of the famous Emperor Constantine I. Apparently, it also hosted the remains of her sister Helena, who died in AD 360, 6 years after Constantina. The Mausoleum does not show evident astronomical alignments but the disposition of the windows in the drum, associated with other significant features of the structure, created visible light phenomena during certain days of the year that were linked to the Saint's feast or holy days.

The identification of these occurrences in the monumental complex and their variations throughout the centuries are the main focus of this study, which is based on the fundamental work by Rasch and Arbeiter (*Das Mausoleum der Constantina in Rom*. Philipp Von Zabern, Mainz Am Rhein, 2007) where reliable plans and architectonic reconstructions are provided. The measures of the orientations were performed on satellite images and on graphic plans, assuming the average value and considering the related standard deviation as uncertainty.

Introduction

Several recent studies have revealed the importance of studying the orientation of ancient Roman monuments that show the existence of intentional light effects in Roman architecture, as for the Basilica Neopitagorica di Porta Maggiore in Rome

F. Carnevale (✉) · M. Monaco
La Sapienza University of Rome, Rome, Italy

G. Magli et al. (eds.), *Archaeoastronomy in the Roman World*, Historical & Cultural Astronomy, https://doi.org/10.1007/978-3-319-97007-3_5

Fig. 5.1 A satellite view of the area of Sant'Agnese fuori le Mura in Rome (after Bing Maps)

(Labianca et al. 2008); the Mausoleo degli Equinozi along Via Appia in Rome (De Franceschini 2012); the Roccabruna and the Accademia in the Villa Adriana (De Franceschini and Veneziano 2013); the Pantheon (De Franceschini 2014; Hannah and Magli 2011); and the Mausoleo di Adriano in Rome (De Franceschini and Veneziano 2015).

For the Mausoleo di Santa Costanza in Rome the light phenomena inside the monument have never been investigated. Our intent is to verify the existence of relationships between the main orientations of the buildings and the illumination of the interiors.

The Mausoleum is located on Via Nomentana, about two miles outside the Aurelian Wall, in the archaeological complex of Sant'Agnese fuori le Mura. Different constructions make up the complex: the catacombs of Sant'Agnese, dated between the second and the fourth centuries AD; the sacellum ad corpus, a votive chapel built over St. Agnes' burial in the first half of the fourth century AD; the horseshoe-shaped cemeterial basilica associated with St. Agnes, and the Mausoleum itself, the last two built in the middle of the fourth century AD by Constantina or Costantia, the eldest daughter of the Roman Emperor Constantine I the Great (Fig. 5.1).

According to traditional literature (Frutaz 1960; Johnson 2009; Krautheimer 1937; Rasch and Arbeiter 2007), the Mausoleum was built in the second quarter of the fourth century, probably between AD 337 and 350, when Constantina was residing in Rome between her two marriages. As reported by the contemporary historian Ammianus (Amm. Marc. XIV, 11, 6; XXI, 1, 5), Constantina died in AD

354 in Bithynia, Asia Minor, and was buried in the mausoleum. Later, the remains of her sister Helena, who died in AD 360, also were buried in it.

More recent excavations have uncovered the foundations of a triconch building under the area originally occupied by the vestibule of the Mausoleum (Stanley 1994). This fact led some authors to postpone the construction of the Mausoleum to the second half of the fourth century AD (Ringbom 2003) or even the late fourth or early fifth century AD (Stanley 1994).

The Mausoleum was connected to the cemeterial basilica through a narthex mid-way along the North side and was originally surrounded by a circular portico of which only the perimeter wall still exists. Originally the only source of light were the drum windows that today are partially occluded in the lower part, given that the small windows of the ambulatory are commonly considered as later additions (Johnson 2009; Rasch and Arbeiter 2007). Soon after the construction, possibly in the late fourth century AD, a rectangular tower (*lucernario*) was added to the drum on the South West side, blocking one of the drum windows (Johnson 2009; Prandi 1942–1943).

The architecture is formed by a central Rotunda, surrounded by a barrel-vaulted ambulatory. The interior was richly decorated. The decoration of the ambulatory vault is preserved although actually heavily restored (see Matthiae 1967) and consists of mosaics with geometric patterns and harvest scenes. The drum was decorated with marble marquetries and the dome with mosaics representing biblical scenes; although destroyed in 1620, those decorations are known thanks to several Renaissance drawings (see Rasch and Arbeiter 2007; Ringbom 2003). Lastly, the ambulatory niches also were decorated with mosaics, which probably disappeared between the fifteenth and the seventeenth centuries. Only the decorations of the niches of the cross-axis are preserved, representing respectively the *Traditio Legis* and the *Traditi Clavium*.

An enormous porphyry sarcophagus with the remains of Constantina was preserved in the Mausoleum (Fig. 5.2). It was moved for the first time in 1467, but was brought back soon after. In 1790 it was finally moved to the Vatican Museums (Johnson 2009) and a cast replica was placed in the mausoleum, in the south-west niche of the ambulatory. Its original position is an open question. According to scholars, there were three possibilities: for Bosio et al. (1659) and Frutaz (1960) the sarcophagus was in the south-west niche; for Prandi (1942–1943) and Johnson (2009) it was under the small tower and placed over the still-existing porphyry oval mounted in the pavement; while for Jubaru (1904) it was in the centre of the mausoleum.

Today the sarcophagus is placed over a modern support and it is over 2.25 m in height (see Fig. 5.2a). Its original dimensions can be inferred from a drawing by Desgodets published for the first time in 1682 (see Desgodets 2008) in which the measurements are reported in Pied de Roy. In a recent study (not yet published) we have determined the correct conversion factor for all of the drawings made by Desgodets:

$$1 \text{ Pied de Roy} = 0.326 \text{ m}.$$

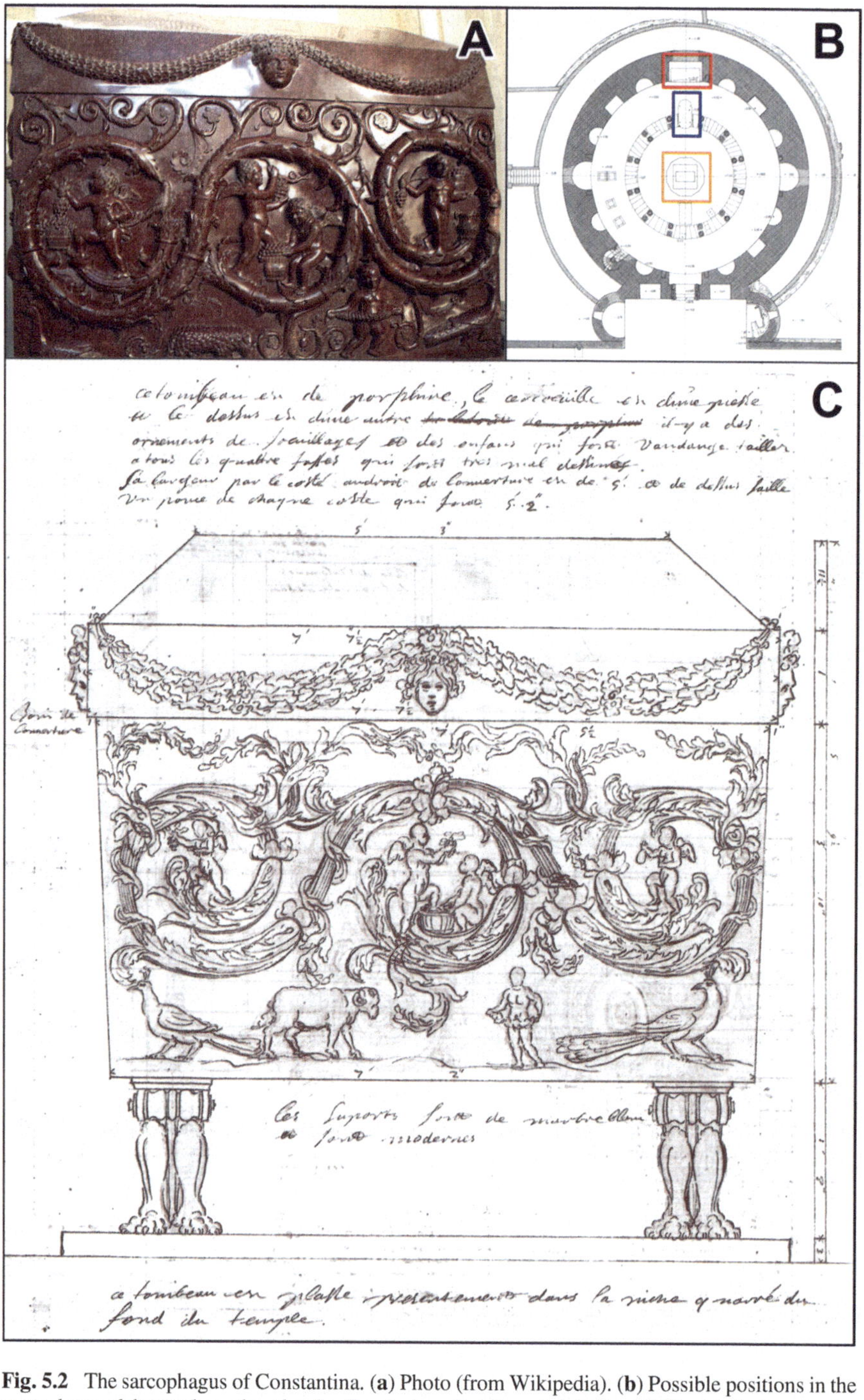

Fig. 5.2 The sarcophagus of Constantina. (**a**) Photo (from Wikipedia). (**b**) Possible positions in the mausoleum elaborated on the plan by Rasch (see Rasch and Arbeiter 2007). (**c**) A drawing by Desgodets (2008); it is worth noting the peculiar shape of the sarcophagus cover

Table 5.1 A chronology of events linked to the Mausoleum

Date	Event
Second–fourth centuries AD	Catacombs.
AD 305	St. Agnes' death.
AD 337–350 AD 337–350	The Paleochristian basilica and Mausoleum was built by Constantina. A sacellum ad corpus was built upon St. Agnes' tomb.
Late fourth century AD	Maybe, the small tower was added to the original Mausoleum, blocking the south-west window.
Fifth century AD	In the *Passio Latina*, Santa Costanza is described as a leper who later was cured by St. Agnes.
AD 625–638	Pope Honorius I built the basilica of Sant'Agnese in place of the sacellum ad corpus.
AD 865	In the *Liber Pontificalis*, the Mausoleum is mentioned as the "Aecclesia Sanctae Constantiae".
1256	Pope Alexander IV consecrated the altar "beate Constantie filie Constantini".
1450	Rucellai describes only one porphyry sarcophagus in the Mausoleum.
1467	Pope Paul II moved the sarcophagus to the Piazza San Marco, but Sixtus IV brought it back to the Mausoleum.
Sixteenth century	Santa Costanza (with Attica and Artemia) is mentioned for the first time in Martyrologies.
1538	In a drawing by Francisco De Hollandia the sarcophagus is shown in the south-west niche.
1567	In a drawing by Peruzzi there is an oval in the centre of the Mausoleum that is compatible with the sarcophagus' dimensions. Peruzzi reports only one sarcophagus, located in the niche in front of the entrance.
1606	A porphyry sarcophagus was moved from Santa Costanza or from Santa Agnese to a chapel in San Pietro to contain the remains of the Saints Simone and Giuda. This sarcophagus is not visible anymore, because it was covered by an altar during the twentieth century.
1620	The mosaics in the dome were replaced by frescos.
1791	Pope Pius VI took the sarcophagus to its current position, in the Vatican Museums.

This allows us to determine the original dimensions of the sarcophagus: length 2.48 m, width 1.63 m and height 1.89 m.

Through the centuries, the Mausoleum and the whole complex underwent several transformations (see Table 5.1). In AD 625–638 Pope Honorius I built the basilica of Sant'Agnese in place of the older sacellum ad corpus. During the Middle Ages the Mausoleum was transformed in the Aecclesia Sanctae Constantiae (church of Santa Costanza), mentioned for the first time in AD 865 by the *Liber Pontificalis* (Frutaz 1960; Ringbom 2003).

Methodological Approach

As already shown in a study on the orientation of the Greek theatres (Monaco et al. 2017), the azimuth can be determined in three different ways: on-site measurements; deriving it from archaeological plans; or measuring it on satellite images. For the Mausoleum of Santa Costanza on-site measurements with a professional compass can be unreliable, because the monument is located in a dense urban area where magnetic measurements can be strongly affected by nearby structures.

The best methodology to derive the orientations from archaeological plans would be that presented by Ranieri (2014), who chose the most reliable plan from among all those available. There are only two plans that appear reliable if we wish to determine the orientation of the different buildings in the complex: the one published by Rasch (Rasch and Arbeiter 2007) and the one edited by Pavolini (Magnani Cianett and Pavolini 2004) (Fig. 5.3). Neither plan indicates whether the north arrow indicates true north or magnetic north. Hence, we decided to use satellite images as well, available for free on the web (*Comune di Roma—Geoportale*, Google Earth and Bing Maps), which present a resolution appropriate to our purposes.

The orientation of each building was determined by calculating the average value, and the standard deviation was considered as the error of the measurement. The bearings were determined by measuring the angle from north to the axis of the buildings in a clock-wise direction (from north through east).

The analysis of the light phenomena for the monuments was done by carefully taking into account the restorations and the invasive changes that affected the architecture structures over time.

We used the software *Stellarium* to simulate the ancient sky and to determine the days of the year when different phenomena occurred.

Orientations

Our measurements of the azimuth for the different buildings composing the archaeological complex are shown in Table 5.2. Five azimuth values (°) are presented:

1. $AZ._B$ measured on satellite image from Bing Maps (2017);
2. $AZ._G$ measured on satellite image from Google Earth (2016);
3. $AZ._C$ measured on ortophoto from *Comune di Roma—Geoportale*;
4. $AZ._P$ measured on the plan edited by Magnani Cianett and Pavolini (2004); and
5. $AZ._R$ measured on the plan published by Rasch and Arbeiter (2007).

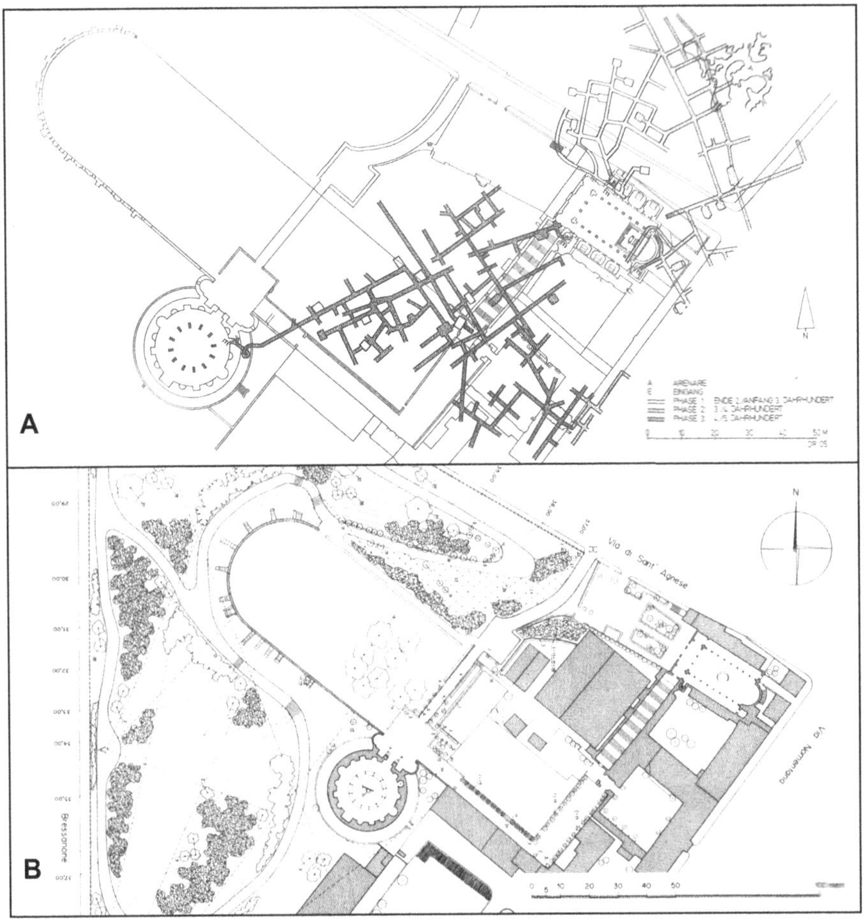

Fig. 5.3 Plans of the archaeological complex. (**a**) After Rasch and Arbeiter (2007); (**b**) after Magnani Cianetti and Pavolini (2004)

The Mausoleum of Constantina

Two different light phenomena were considered:

1. one connected with the main axis and related to the window in front of the entrance; and
2. one connected with the possible position of the sarcophagus and related to the drum windows.

The astronomical values for the Mausoleum are reported in Table 5.3 while the periods of illumination of the possible locations of the sarcophagus in the interior are reported in Table 5.4.

Table 5.2 Azimuth values measured for the different buildings comprising the complex

ID	Edifice	Date AD	AZ_B		AZ_G		AZ_C		AZ_P		AZ_R		Average		St. Dev.	
			°	′	°	′	°	′	°	′	°	′	°	′	°	′
1	Mausoleum	~450	40	16	39	58	39	59	38	36	39	48	**39**	**43**	0	39
2	Cem. Basilica	~450	130	16	129	58	129	59	128	36	129	48	**129**	**43**	0	39
3	Hon. Basilica	638	304	30	303	46	305	52	302	08	303	49	**304**	**0**	1	24

KEY: *Mausoleum* Mausoleum of Constantina, *Cem. Basilica* Constantinian Cemetarial Basilica, *Hon. Basilica* Honorian Basilica, AZ_B Azimuth measured on Bing Maps (2017), AZ_G Azimuth measured on Google Earth (2016), AZ_C Azimuth measured on *Comune di Roma—Geoportale* (2012), AZ_P Azimuth measured on the plan by Magnani Cianett and Pavolini (2004), AZ_R Azimuth measured on the plan by Rasch and Arbeiter (2007), *St. Dev.* Standard Deviation
The average values are shown in bold print

Table 5.3 Azimuth, altitude, declination and the corresponding date in AD 350 of the possible heliophany occurring from the window in front of the entrance

Event	Azimuth	Altitude	Declination	Corresponding date in AD 350
Start of the heliophany	217° 13′	24° 54′	−14° 53′	February 08
End of the heliophany	222° 08′	29° 20′	−08° 54′	February 25

Table 5.4 The period of illumination of the sarcophagus depending on its position

Position	Period of illumination
Centre	March 29–September 13
Porphyry oval	May 1–October 15

Considering the azimuth of the main axis, the altitude and the declination (together with the position and the dimension of the window in front of the entrance) we have determined that in AD 350 the heliophany occurred between February 8th and 25th (see Fig. 5.4a).

Regarding the phenomena connected with the position of the sarcophagus in the interior, we have determined that the centre of the building was illuminated between 29 March and 13 September (see Fig. 5.4b), while the area beneath the small tower, where the porphyry oval is mounted in the pavement, was illuminated between 1 May and 15 October (see Fig. 5.4c).

The Constantinian Basilica

The astronomical values for the Constantinian Basilica are shown in Table 5.5, and a reconstruction of it is shown in Fig. 5.5.

No evident solar alignment is suggested by this azimuth value. Due to the bad state of conservation, it is not possible to perform reliable studies of light phenomena in the Basilica.

The Honorian Basilica

The astronomical values for the Honorian Basilica are shown in Table 5.6, and a reconstruction of it is shown in Fig. 5.6.

An alignment toward the Summer Solstice Sunset is clearly indicated by the azimuth value. Due to the numerous and important restorations made through the centuries, no further study of light phenomena was attempted.

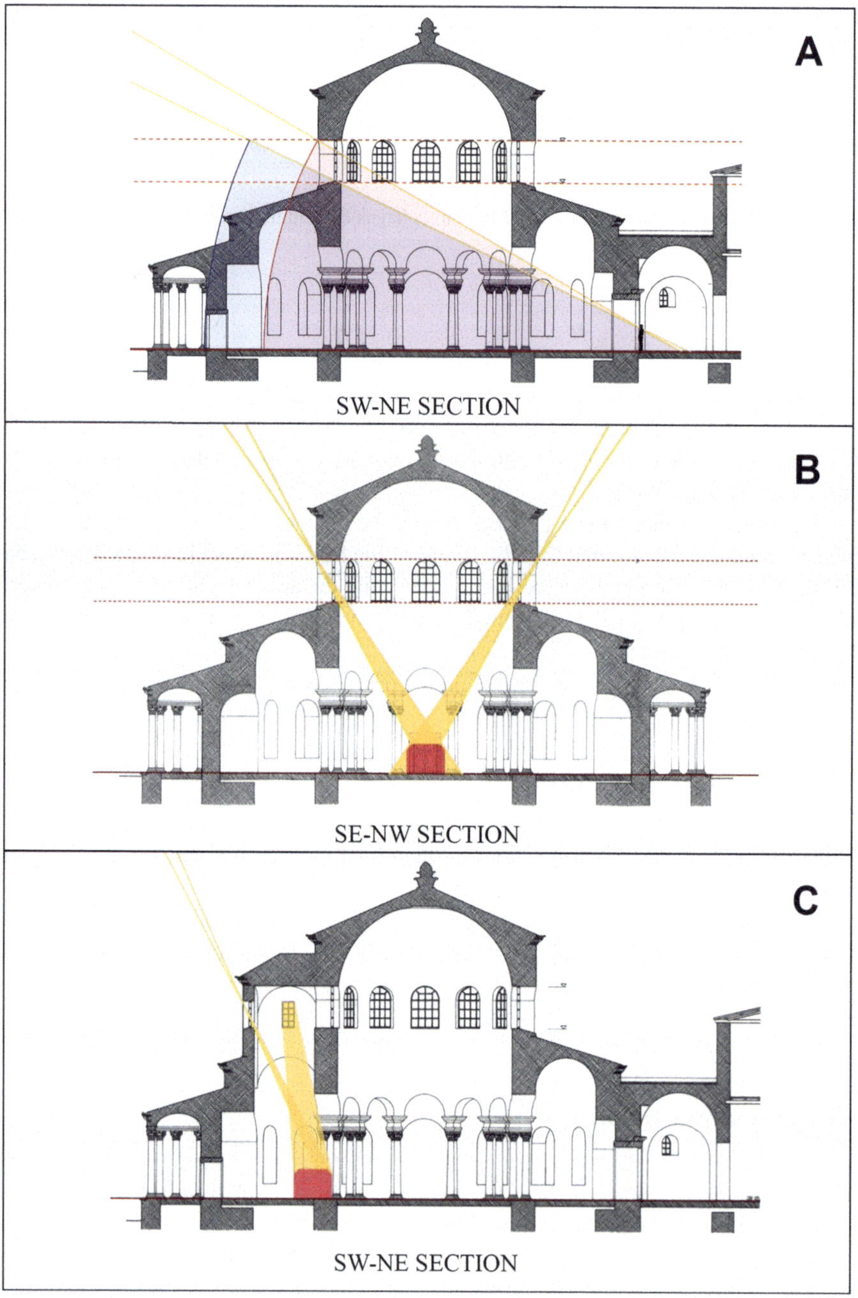

Fig. 5.4 The mausoleum of Constantina. (**a**) Representation of the minimum and maximum altitude of the ray illuminating the entrance (see Table 5.3). (**b**) Representation of the minimum and maximum altitude of the ray illuminating the centre (see Table 5.4). (**c**) Representation of the minimum and maximum altitude of the ray illuminating the porphyry oval in its current position (see Table 5.4) (modified from Rasch and Arbeiter 2007)

Table 5.5 Azimuth, altitude and declination of the Constantinian Basilica

	Azimuth	Altitude	Declination
Constantinian Basilica	129° 43′	10° 14′	−20° 31′

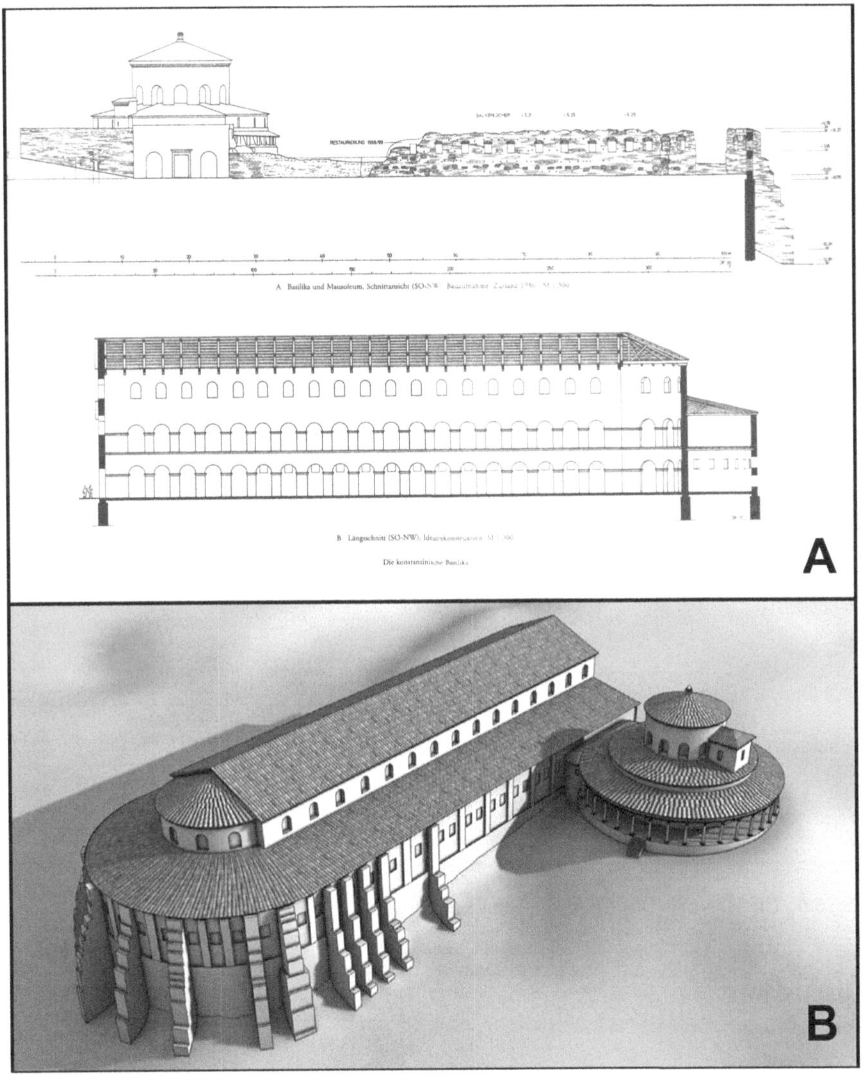

Fig. 5.5 Reconstructions of the Constantinian Basilica of Sant'Agnese. (**a**) Drawing (after Rasch and Arbeiter 2007). (**b**) 3D reconstruction (after Bardill 2012)

Table 5.6 Azimuth, altitude and declination of the Honorian basilica and of the Summer Solstice sunset

	Azimuth	Altitude	Declination
Honorian Basilica	304° 0′	0° 36′	+24° 42′
Summer Solstice sunset	303° 13′	–	+23° 39′

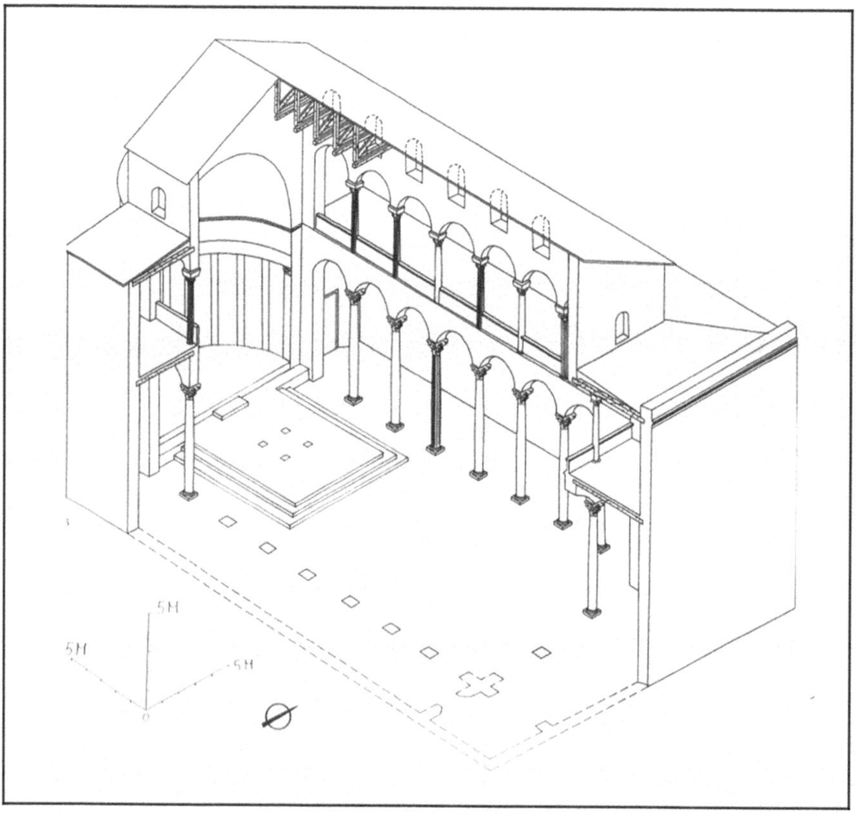

Fig. 5.6 The Honorian Basilica (from Brandenburg 2004)

Discussion

Two different light phenomena occurred in the Mausoleum: (1) one relating to the entrance and the window in front of it; and (2) those relating to the positions of the sarcophagus.

Prandi (1942–1943), Rasch (Rasch and Arbeiter 2007) and Johnson (2009) suggested that the small tower or *lucernario* is an early addition in respect to the old drum of the Mausoleum: this means that between the foundation and construction of the *lucernario*, the light could directly hit the entrance each year from 8 to

25 February. These days corresponded to the *Parentalia*, a pagan feast celebrating the family dead. Once the *lucernario* was built, that window was plugged and the light phenomenon was interrupted. This kind of heliophany resembles the one occurring in the Pantheon on 21 April (the founding day of Rome): in both cases the light appears to be an integral part of the celebration of the festivity.

The Mausoleum has a peculiar feature: the maximum inclination of the sunlight passing through the windows ends at the centre of the Mausoleum: this is perfect for illuminating an object placed in that position. According to many archaeologists, the sarcophagus of Constantina was located in the large niche in front of the entrance from the early days. This niche never receives direct light, so we find it a weird location for a red sculpted porphyry sarcophagus, a material known for its reflection properties, with that particular shape of the cover.

According to tradition (see Frutaz 1960), from AD 360 to 1606 two sarcophagi were conserved inside the Mausoleum: one belonging to Constantina and the second to her sister, Helena. Probably in the beginning (see Ringbom 2003) the first one was the only one present in the monument, and the arrival of Helena's may have forced the finding of another location for it. The presence of two sarcophagi imply two different positions inside the Mausoleum, but we could not find citations about this in literature. Nonetheless, a drawing by Peruzzi (see Rasch and Arbeiter 2007) shows an oval at the centre of the Mausoleum with a size compatible with the sarcophagus dimensions. A porphyry oval is now present and mounted in the pavement under the *lucernario* (his width corresponds perfectly to the width of the sarcophagus). We have no certainty whether the ovals were two or one (the central), which was moved eventually under the *lucernario*.

However, we believe that the two positions are certain. As a matter of fact, the light phenomena still occurring in the building insist precisely on these two positions.

On 1606 a sarcophagus (probably that of Helena) was taken from the complex and moved to San Pietro to contain the remains of Simone and Giuda. Its current arrangement (behind a more recent altar) does not allow the study of it. The centre of the building was used for the current altar that we can still admire.

Later, in 1791, the porphyry sarcophagus of Constantina was permanently moved to the Vatican Museum.

Archaeoastronomical features are also present in the more recent Honorian Basilica, since its entrance is oriented toward Summer Solstice sunset. Until the seventeenth century the Basilica had the entrance at the same level of the *matroneo*, which was several meters above the pavement level. When the sunlight penetrated from the entrance, it illuminated the golden mosaic in the apse depicting Sant'Agnese and Pope Honorius.

Due to its bad state of conservation, the alignment of the Constantinian Basilica of Sant'Agnese cannot be discussed.

Conclusions

This study adds new evidence that archeoastronomy can help in clarifying archaeological questions, as several recent publications have shown.

We have found that:

1. The entrance of the Mausoleum of Constantina was illuminated between 8 and 25 February, days corresponding to the Festival of Parentalia in honour of family ancestors.
2. The identified light phenomena endorse the theory that a sarcophagus could originally have been placed in the centre, as happened with the burial chambers of imperial mausolea. The peculiar shape and material of the sarcophagus of Constantina would allow an incredible reflection effect of the light.
3. In Late Antiquity or the Early Middle Ages, when the Mausoleum was transformed into a church, the sarcophagus was possibly moved to under the *lucernario*, which corresponds to the porphyry oval that is still mounted on the pavement.
4. Later, the sarcophagus was moved in the south-west niche, as testified by some fifteenth century drawings. In this position it could not be illuminated by natural light.
5. Like several important monuments of antiquity, the Honorian Basilica shows an alignment toward the Summer Solstice sunset.

Acknowledgements We would like to thank Marcello Ranieri (INAF-IAPS Roma) for review work and valuable help.

References

Ammianus Marcellinus, XIV, 11, 6; XXI, 1, 5. Available via DIALOG. http://penelope.uchicago.edu/Thayer/e/roman/texts/ammian/home.html. Accessed 30 June 2017.

Bardill, J. (2012). *Constantine, Divine Emperor of the Christian Golden Age*. New York: Cambridge University Press.

Bosio, A., Severano, G., & Aringhi, P. (1659). *Roma Subterranea Novissima*. Collegium Graphicum, Colonia. Available via DIALOG. http://digi.ub.uni-heidelberg.de/diglit/bosio1659bd2/0095. Accessed 30 June 2017.

Brandenburg, H. (2004). *La Basilica di S. Agnese e il Mausoleo di Costantina Augusta (S. Costanza)*. Milano: Jaca Book.

De Franceschini, M. (2012). Orientamenti equinoziali nel Mediterraneo: Il tempio preistorico di Mnajdra a Malta e il Mausoleo degli Equinozi a Roma. *XIV Seminario Associazione Ligure per lo Sviluppo degli Studi Archeoastronomici*, 18–29.

De Franceschini, M. (2014). Il Pantheon di Roma: Nuove immagini dei fenomeni luminosi. L'arco di luce. *XVI Seminario Associazione Ligure per lo Sviluppo degli Studi Archeoastronomici*, 129–140.

De Franceschini, M., & Veneziano, G. (2013). Architecture and archaeoastronomy in Hadrian's Villa near Tivoli, Rome. *Nexus, 15*(3), 457–485.

De Franceschini, M., & Veneziano, G. (2015). Archeoastronomia nel Mausoleo di Adriano Castel Sant'Angelo (Roma). *XVII Seminario Associazione Ligure per lo Sviluppo degli Studi Archeoastronomici*, 5–17.

Desgodets, A. (2008). *Les Édifices Antiques de Rome*. Édition fac-similé du manuscrit inèdit 2718 de l'Institut de France avec transcriptions, annotations et reproduction des planches de l'ouvrage publié en 1682.par L. Cellauro & G. Richaud. Roma: De Luca Editore d'Arte.

Frutaz, A. P. (1960). *Il Complesso Monumentale di S. Agnese e S. Costanza*. Città del Vaticano: Tipografia Poliglotta Vaticana.

Hannah, R., & Magli, G. (2011). The role of the sun in the Pantheon's design and meaning. *Numen, 58*(4), 486–513.

Johnson, M. J. (2009). Mausoleum of Constantine, Rome. In M. J. Johnson (Ed.), *The Roman imperial mausoleum in late antiquity* (pp. 139–156). New York: Cambridge University Press.

Jubaru, F. (1904). La decorazione bacchica del mausoleo cristiano di S. Costanza. *L'Arte, 7*, 457–468.

Krautheimer, R. (1937). *Corpus Basilicarum Christianarum Romae (IV–IX cent.)* (Vol. I, Number I, pp. 14–38). Roma: Citta del Vaticano.

Labianca, L., Sciortino, I., Gaudenzi, S., Patané, A., Polcaro, V. F., & Ranieri, M. (2008). An archaeoastronomical study of the neo-Pythagorean basilica at Porta Maggiore in Rome. In *Proceedings of the SEAC 2007 and Oxford VIII conference astronomy and cosmology in folk tradition and cultural heritage* (pp. 195–199). Archaelogia Baltica 10, Klaipeda University Press.

Magnani Cianett, M., & Pavolini, C. (Eds.). (2004). *La Basilica Costantiniana di Sant'Agnese: Lavori Archeologici e di Restauro*. Milano: Electa.

Matthiae, G. (1967). *Mosaici Medioevali delle Chiese di Roma* (pp. 3–53). Roma: Istituto Poligrafico dello Stato.

Monaco, M., Carnevale, F., & Ranieri, M. (2017). A study on the orientation of Greek theatres. In A. Orlando (Ed.), *The light, the stones and the sacred* (pp. 107–121). Cham: Springer (Astrophysics and Space Science Proceedings, Volume 48).

Prandi, A. (1942–1943). Osservazioni su S. Costanza. *Rendiconti della Pontificia Accademia Romana di Archeologia, 19*, 281–304.

Ranieri, M. (2014). Digging the archives: The orientation of Greek temples and their diagonals. *Mediterranean Archaeology & Archaeometry, 14*(3), 15–27.

Rasch, J. J., & Arbeiter, A. (2007). *Das Mausoleum der Constantina in Rom*. Mainz Am Rhein: Philipp Von Zabern.

Ringbom, Å. (2003). Dolphins and mortar dating. Santa Costanza reconsidered. In Å. Ringbom & R. Suominen-Kokkonen (Eds.), *Songs of Ossian. Festschrift in Honour of Professor Bo Ossian Lindberg* (pp. 23–42). Vaasa: Taidehistorian Seura.

Stanley, D. J. (1994). New discoveries at Santa Costanza. *Dumbarton Oaks Papers, 48*, 257–261.

Part III
Roman Towns

Chapter 6
Establishing a New Order: The Orientation of Roman Towns Built in the Age of Augustus

A. César González-García, Andrea Rodríguez-Antón, David Espinosa-Espinosa, Marco V. García Quintela, and Juan Belmonte Aviles

Abstract Urbanism in most areas of Western Europe occurred at the time of the Roman Empire when several hundred new towns were founded, notably under Augustus. Those towns were planned to incorporate astronomical phenomena as images of propaganda of their rulers, or to connect the city to the gods. The visual effect of the Sun rising in line with the orientation of the city at a given moment in its yearly movement was thus sought and incorporated for its ritual meaning. Special moments allegedly related to Augustus were considered, in particular Winter Solstice and Autumn Equinox.

Introduction

Phenomena like the 'Manhattanhenge' have a great visual impact for people today, despite being due to mere chance alignment of an orthogonal grid of streets that happens to face sunset twice a year. However, urban city planning that accommodated celestial movements was not so strange in certain societies, like the Chinese for instance (see Pankenier 2014). Urban grids from classical antiquity, and the Roman realm in particular, may also have incorporated a ritual meaning that connected their layout with the sky.

A. C. González-García (✉)
Instituto de Ciencias del Patrimonio, Incipit-CSIC, Santiago de Compostela, Spain
e-mail: a.cesar.gonzalez-garcia@incipit.csic.es

A. Rodríguez-Antón · J. B. Aviles
Instituto de Astrofísica de Canarias & Universidad de La Laguna, La Laguna, Spain
e-mail: ara_ext@iac.es; jba@iac.es

D. Espinosa-Espinosa · M. V. García Quintela
Departamento de Historia, Faculdade de Xeografía e Historia, Universidade de Santiago de Compostela, Santiago de Compostela, Spain
e-mail: david.espinosa@usc.es; marco.garcia.quintela@usc.es

G. Magli et al. (eds.), *Archaeoastronomy in the Roman World*, Historical & Cultural Astronomy, https://doi.org/10.1007/978-3-319-97007-3_6

The foundation of new cities was a key element in Roman expansion, especially in the western part of the Empire (Laurence et al. 2011) where literally hundreds of towns were founded—or re-founded over pre-existing settlements—from the third century BCE to the fourth century CE. Laurence et al. (2011: 138) state that "The grid of streets . . . must be analysed as a key aspect of Roman urbanism."

These towns were built in a characteristic shape, incorporating a grid of orthogonal streets with minor variations (Gros and Torelli 1994; Kaiser 2011). The Roman towns show a high degree of structure and standardization, particularly in the west of the Roman world (Laurence et al. 2011).

The founding of cities in antiquity was not an issue that was left to the mercy of chance, nor was it done in empty spaces devoid of external components. The founding of a town had a deep political and ritual meaning (Woodward and Woodward 2004). In many cases, particularly relating to Roman colonies, it had a clear practical dimension of land control, but also a highly symbolic one, as the new settlement highlighted the role of Rome as the ruler city while the new town was viewed as a small portion of the *Urbs* itself in the area to be controlled (*quasi effigies parvae simulacraque Romae*).[1]

In practical terms, the local topography (i.e., rivers, hills, mountains, pathways) affected the location of new settlements and perhaps also their orientation. In other words, cities accommodated and incorporated the surrounding landscape as the *Agrimensores* and other ancient sources state, such as Hyginio Gromatico (Hyg. Grom., La. 168.3–5). What is generally less taken into account is whether these towns, in one way or another incorporated celestial objects or events as part of that landscape (see e.g. González-García and Magli 2014).

In symbolic terms, numerous sources document how oracles had to be consulted, or rituals had to be elaborated to ascertain the suitability of the time, place or settlers (Rykwert 1988), and a ritual to consecrate the space was a prerequisite, at least until the early centuries of the Roman Empire (Briquel 2008). In many cases such a foundation also involved spatial planning and its delimitation or partitioning.

How Roman cities were orientated as part of these rituals is still a matter of intense debate. The classical sources point to different methods, either purely practical, although based in astronomical observations (such as Vitruvius in *de Architectura* I, 6) or possibly by a more ritual use of astronomical phenomena, particularly sunrise or sunset, as implied by the accounts of the *Agrimensores* (e.g. Frontinus *De limit., 10.20–11.6 Th; 11.9–14 Th*; see Espinosa-Espinosa and González-García 2017 and González-García and Magli 2014 for recent reviews).

The last century of the Republic showed a clear impulse for the building of monumental structures in Rome and in the provinces, that increased during Julian and Augustan times. In fact, a substantial part of the remains visible today in the main area of the forum in Rome date to this period. Likewise, in many towns in the West, massive building plans and embellishment were carried out during the time of Augustus and his successors. For example, an Augustan colony replaced the old

[1]Gel. XVI, 13, 9.

town in *Brixia* (present day Brescia, e.g. L'Année Épigraphique 1996: 726). The series of new colonies founded at the time of Augustus form a homogeneous group with an extremely standard design (González and Ruiz de Arbulo 2010; Owens 1992) and they are the main focus of this paper.

It has been argued that Augustus used astrological images as part of his propaganda agenda, perhaps also reflecting a Hellenistic tradition of using the sky to enhance the ruler cult (see Ferro and Magli 2012 for the connection between the orientation of the city grid in Alexandria and astronomy, and Belmonte and González-García 2010 for the use of astronomy at the *Hierothesion* of Antiochus of Commagene). This is the case with the appearance of the so-called Caesar's Comet in 44 BCE: Octavianus used such an appearance to highlight his role as sole heir of the *Divus Iulius* (Ramsey and Licht 1997). Suetonius (*Aug* XCIV, 12) mentions that Augustus' horoscope was under the sign of Capricorn, something Augustus himself proclaimed and published despite the fact that he was born nearly at the Autumn Equinox, with the Sun in Libra (see e.g. Barton 1995). As part of this propaganda, a large number of coins were struck with a Capricorn sign in its reverse at the time of Augustus and his successors. The Capricorn sign appeared in architectural decorations, in gems (such as the *Gemma Augustea*) and glass pastes (Galinsky 2012). Capricorn was the constellation housing the Sun during Winter Solstice, the time of solar reborn that Augustus used as an allegory for the re-foundation of Roman customs he was advocating (Barton 1995). Astronomical images and orientations have also been identified in several monuments erected at his time, such as the Pantheon or the *Horologium Augusti* in Rome (Hannah and Magli 2011; Haselberger 2014; Rehak 2006).

Our Hypothesis

Roman town planning during the Age of Augustus had a strong component of propaganda to impose the image of the ruler as warrant of peace and restorer of the order (Espinosa-Espinosa and González-García 2017). The use of astronomical imagery played a key role in this program.

Thus, our research question is: To what extend did such a program of propaganda extend also to the way Roman settlements in the West were laid out to incorporate orientations towards particular directions, such as Winter Solstice sunrise?

Our working hypothesis then is that the principal public buildings, especially the Forum, or the main axes of the city, faced towards particular points on the horizon where sunrise or sunset at certain times of the year connected to Augustus could be spotted. In particular, the Winter Solstice seems to have been the main target according to our hypothesis. A secondary target would be the equinox when the *dies natalis Augusti* was supposed to happen.

This hypothesis has already been tested in a small number of Roman towns, with positive results. In recent years there have been two main efforts to understand the

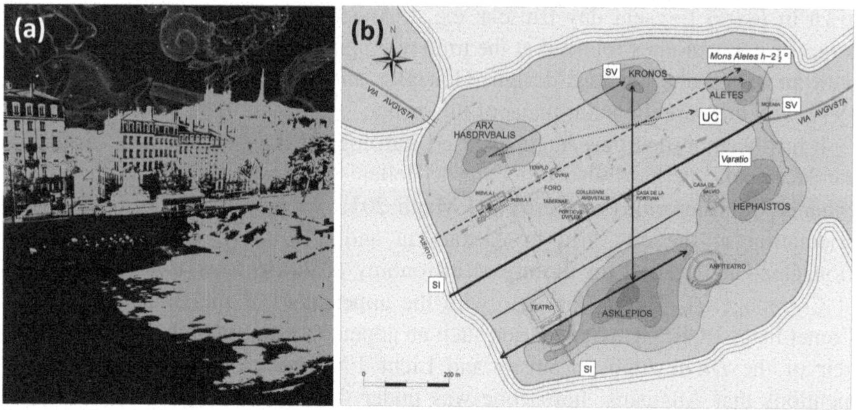

Fig. 6.1 (**a**) The Roman amphitheater of Lyon was associated with the altar of the *Tres Galliae* where a yearly meeting of all Gaullic tribes was held on 1 August. On that date, at dawn the constellation of Capricorn could be seen setting behind the municipal temple of the Imperial cult. (**b**) The urban layout of Carthago Nova surely incorporated the Punic design, but the location of the Forum and its temple to Rome and Augustus was chosen so that Summer Solstice sunrise happened on top of Mons Aletes, which was arguably devoted to the mythical founder of the town

issues at stake, either by focusing on particular case studies or by trying to collect samples of data to verify if a common pattern exists.

For Italy and Spain, Magli (2008) and González-García et al. (2014) have shown that a fair number of cities have orientations consistent with sunrise on the Winter Solstice. Although the sample is far from being complete, these works do seem to point towards the actual use of solar observations in the orientation of Roman towns, and the Winter Solstice sunrise seems to have been particularly important.

It is fascinating that a fair number of these towns with astronomical orientations either were founded or reconstructed at the time of Augustus. Three cases of interest where particular astronomical relations appear are *Lugdunum* (Lyon) in Gaul, Aosta (ancient *Augusta Pretoria Salassorum*) in Italy and *Carthago Nova* (Cartagena) in Spain.

Lugdunum presents an elaborate case where the urban layout fully conforms to Roman models but also matches Gallic traditions (García Quintela and González-García 2014; González-García and García Quintela 2014). In particular, the orientation of the town, which is clearly a Roman colony, could be related to a traditional Gallic festival celebrated on 1 August nearly at the foundation period on 43 BC, 31 years before establishing the altar for the Imperial cult, and the festivities to be celebrated on that same day. During that period, the proposed location of the altar of the *Tres Galliae*, and the orientation of the amphitheater associated with it, allowed viewing of the setting of the constellation of Capricorn over the municipal temple of the Imperial cult built in the heart of the Roman city (Fig. 6.1a). This was a highly evocative game of legitimation of Imperial power that linked the Gaulic festival, performed to honor Rome, and a symbol of that power (Capricorn) associated with the sanctuary of the Imperial cult (González-García and García Quintela 2014).

In *Augusta Pretoria* we find a similar pattern with orientations towards the Winter Solstice. At Mediterranean latitudes, where the direction of the *decumanus* generally agrees with the course of the Sun, the perpendicular street, *cardo*, would be out of this solar arc. However, for Aosta, Bertarione and Magli (2015) have recently revealed that due to its location in a deep valley within the Alps, the south end of the *cardo* is, also, inside the solar range. That part of the city is facing the top of a nearby mountain, allowing this street to be oriented to the Winter Solstice sunrise, the moment when the Sun was in the constellation of Capricorn.

Carthago Nova (see Belmonte et al. 2016; González-García et al. 2015) suffered a re-foundation of the older Carthaginian town with a renewed reading of the local landscape under the Roman and particularly within the Augustan frame of mind. The original Punic layout was reconsidered and, for instance, the area of the forum of the new colony had a slightly modified orientation, such that the temple of the Imperial cult was oriented towards the Summer Solstice sunrise on top of a nearby hill devoted to a mythical founder of the site (*Mons Aletes* according to Polybius[2]; Fig. 6.1b). Augustus or his allies in *Carthago Nova* linked the new founder with the old one. The landscape of Punic times was modified, reinterpreted and changed along with progressive Romanisation.

It is clear that the two types of studies, survey-like and in-depth, complement each other by providing a detailed picture of the role of astronomy in the orientation and location of cities and public buildings in Roman towns.

Methodology

Thus, we have measured the orientation of a number of cities with archaeologically and epigraphically attested Augustan foundations or reforms carried out at that period, in the ancient Roman provinces of *Hispania*, *Gallia*, *Germania*, *Italia* and North Africa (see Table 6.1).

The methodology employed involves the measurement of those archaeological remains that may give us information about the orientation of the urban planning or in those public buildings erected at the time of Augustus. After determining the direction from which we want to set the orientation, we take the azimuth and the horizon altitude for such directions using tandems, including a precision compass and a clinometer. The error of an individual measure, judging from the scale of the instrument (the Suunto 360PC/360R and the Silva Surveymaster) is $\pm\frac{1}{4}°$ for the azimuth and $\pm\frac{1}{2}°$ for the height of the horizon. Note that in several instances individual measurements will have larger errors due to, above all, the condition of some of the measured remains. Since the instruments used were magnetic, we had to correct the magnetic declination readings. These values are commonly estimated by comparing readings for conspicuous landmarks that can be later be compared with

[2]Pol. 10, 10, 6.

Table 6.1 Augustan towns measured and considered in the text

Roman site	Present name	φ	Ae	he	Aw	hw	δe	δw
Germania								
Ara Ubi/Colonia	Köln	50.93	90½	0½	270½	0.5	0.07	0.25
Gallia Belgica								
Augusta Trev.	Trier	49.75	106	4½	286	5½	–7	14.25
Petrisberg			131	1¾	311	1¼	−23.92	25.81
Augusta Raurica	Augst	47.5	55	1¾	235	2¾	23.93	−21
Vesontio	Besançon	47.25	135	7	315	1½	−22.9	29.64
Aventicum	Avenches	46.86	128	3	308	2	−22.7	26.25
Gallia Lugdunensis								
Augustodunum	Autun	47.00	114½	3¼	295½	1.5	−14.2	17.6
			54½	1	234½	a	23.80	−23.79
Lugdunum	Lyon	45.75	63	0½	243	a	18.48	−18.9
F. Segusiavorum	Feurs	45.7	80	1½	260	1½	7.8	−6.1
Gallia Aquitania								
Argentomagus		46.6	94¾	1	274¾	1	−2.83	3.74
Augustonemeton	Clermont-Ferrand	45.75	90¾	1	270¾	4¼	−0.098	3.80
Med. Santonum	Saintes	45.73	97¾	1	277¾	1	−4.97	5.83
Vesunna	Périgueux	45.17	90	3	270	0[b]	1.95	−0.41
L. Convenarum	Saint Bertrand C.	43.00	118½	4½	298½	3	−17.25	22.42
Gallia Narbonensis								
Vienna	Vienne	45.5	98¾	7	278¾	5	−1.2	9.6
Alba Helviorum	Alba-la-Romaine	44.55	91¼	5½	271¼	5	3	4.2
Arausio	Orange	44.13	76¾	2	256¾	1	10.6	−9
Lugdunum	Laudun	44.11	119¾	1	299¾	0	−20.4	20.4
Nemausus	Nimes	43.80	77¼	0¼	257¼	2	9	−8
Apta Iulia Vulg.	Apt	43.87	95	2¾[b]	275	0¾[b]	−1.8	3.8
Glanum		43.77	81¼	9[b]	261¼	6	12.5	−2.2

Arelate	Arles	43.67	90	0	270	0	−0.4	−0.4
F. Iulii	Fréjus	43.43	54¾	2[b]	234¾	2[b]	26.1	−23.5
Ruscino	Perpignan	42.70	91¾	0	271¾	1¼	−1.68	2.25
Italia								
Tridentum	Trento	46.06	82	8¼	262	13¾	11.43	4.42
Aug. Pretoria	Aosta	45.74	68	10	248	5	22.4	−11.55
Brixia	Brescia	45.54	98	5¾	278	0.1	−1.45	5.27
Verona	Verona	45.44	54	2½	234	0½	26.09	−24.43
Aug. Taurinorum	Torino	45.07	115	2.3	295	3½	−15.86	19.86
A. Bagiennorum	Bene Vagienna	44.56	126	1½	306	1¾	−23.87	25.81
Grumentum	Grumento	40.28	127	3½	307	1½	−24.91	28.17
Hispania Citerior Tarraconensis								
Lucus Aug	Lugo	43.0	71	1¼[b]	251	1¾[b]	14.4	−13.6
Iuliobriga	Retortillo	42.99	124¼	0½	304¼	0	−23.9	23.9
Pisoraca	Herrera	42.59	116½	0	296½	0½[b]	−19.1	19.12
Legio	León	42.6	69¾	1[b]	249¾	0½[b]	15.0	−14.8
Asturica Aug	Astorga	42.46	70½	0½[b]	250½	2	14.2	−14.6
Tarraca	Los Bañales	42.28	50¼	1	230¼	−0¼	28.68	−28.90
Petavonium		42.08	120½	0	300½	0½	−22.1	22.06
Cesaraugusta	Zaragoza	41.65	124¼	0[b]	304¼	0[b]	−25.30	24.44
Bracara Aug	Braga	41.5	72¼	2¼	252¼	0	14.8	−13.6
Barcino	Barcelona	41.38	126¾	2	306¾	4	−25.42	29.46
Aug Bilbilis	Calatayud	41.36	76	3	256	1	12.31	−10.06
Termes	Tiermes	41.33	78	0½	258	1	9.0	−8.6
Tarraco	Tarragona	41.11	119½	−0¼[b]	299½	1¾[b]	−22.4	22.56
Ercavica		40.4	80½	0	260½	0½	7.22	−7.6
Segobriga		39.88	75¾	7½	255¾	0½	15.7	−10.9

(continued)

Table 6.1 (continued)

Roman site	Present name	φ	Ae	he	Aw	hw	δe	δw
Valeria	Las Valeras	39.8	120¾	−0¼	300½	1	−23.3	23.21
Saguntum	Sagunto	39.67	83½	5[b]	263½	4	7.80	−2.88
Libisosa	Lezuza	38.95	87¼	−0½	267¼	0	1.38	−2.5
Lucentum	Alicante	38.3	125¼	0	306¾	2½	−26.9	29.28
Carthago Nova	Cartagena	37.62	61½	2½	241½	0[b]	23.75	−22.5
Hispania Ulterior Betica								
Corduba	Córdoba	37.88	59	0½	239	0	23.99	−24.37
Ituci	Torreparedones	37.75	83	−0½	263	−0½	4.25	−7.63
Lusitania								
Aug Emerita	Mérida	38.9	52¾	0	232¾	1	28.1	−27.9
L.Iulia	Ebora	38.57	58	0	238	0	24.5	−24.9
Pax Iulia	Beja	38.01	127½	−0¼	307½	−0½	−29.3	28.00
Africa Proconsularis								
Utica		37.05	105	0¼[b]	285	1¼[b]	−12.1	12.5
Carthago		36.88	120	0	300	0¼	−23.95	23.4
Uthina		36.61	55½	0.5[b]	235½	0½[b]	27.1	−27.1
Simitthus		36.49	46	0	226	2½[b]	33.6	−32.6
Thuburbo Majus		36.4	126	2	306	3¼	−27.1	30.2
Lepcis Magna		32.6	54½	0	234½	1	29	−29.0
Mauretania Tingitana								
Banasa		34.60	128½	0[b]	308½	0[b]	−31.2	30.4

All sites were measured in situ, and several structures were measured to obtain a mean azimuth. The columns give the Roman Province and the name of the Roman site, its present name, the latitude (φ), the azimuth towards east, the altitude of the horizon towards that direction (he), the azimuth towards west, and its altitude of the horizon, and finally the astronomical declination towards east and west. When considering sites at widely different latitudes, the azimuth comparison renders useless, so it is wise to provide a magnitude independent of the observers' location. Such is the declination, an astronomical magnitude that is also directly comparable with the solar positions. For the time of Augustus, the solar range of declinations varied between the nearly −24° for the Winter Solstice to the 24° of the Summer Solstice, while the equinoxes happened for $\delta = 0°$

[a]Indicates blocked horizon. Corresponding declinations have been calculated assuming an altitude of the horizon of 0°

[b]Denotes that the calculations where performed using a DTM, see text for details

the real directions to correct also for systematic errors in the measuring devices. Where such readings were not possible the magnetic declinations were estimated for the fieldwork dates from the models available in http://www.ngdc.noaa.gov. The values provided in Table 6.1 have been corrected for these systematics and are the mean of a number of measurements, in all cases larger than five. The standard deviation of those measurements are normally smaller than the errors provided above.

The data could then be compared with estimates for the celestial objects visible on that section of the horizon. To perform this comparison our measurements were translated into astronomical declination, resulting in an error estimate of around ¾°, or less than one day in the case of sunrise/sunset.

On a number of occasions, especially for Roman towns in currently sensitive areas due to geostrategic reasons, we could not perform fieldwork and we opted to use satellite images and digital terrain models (DTM). The digital images were normally obtained through the repository Google Earth (GE) and the DTM were accessed via the web application www.heywhatsthat.com (HWT). The accuracy of such measurements depended a lot on a number of considerations (Rodríguez-Antón et al. 2017) and the error estimates on declinations for such measurements had to be determined individually. In general the error estimate was nearly $\pm 1\frac{1}{2}°$ although on occasions it could be larger. These measurements were treated separately in the analysis below, and are indicated with an asterisk in Table 6.1. These data also include the measurements by Magli (2008) in Italy. His database was prepared after measuring the orientation of towns on maps and thus was only based on azimuth data. We have checked each measurement there with GE and we have included the altitude of the horizon computed with HWT.

In the following sections, we treat the data collectively by using histograms to analyse the possibility of finding concentrations of orientations that are not expected according to the null hypothesis that Roman towns were not oriented according to the Sun. In particular, we have used a probability density function with a Gaussian kernel instead of the customary box histogram. This representation allows us to treat each measurement independently with its own error as the width of the Gaussian kernel. Thus, those measurements obtained directly by fieldwork will have a smaller spread and thus are weighted more than those obtained through GE and HWT.

Results

Our database of 64 Augustan towns is drawn from the orientations of 181 Roman towns in *Hispania*, Italy and parts of North Africa, France, Germany, and Switzerland measured mainly by our team during the last decade (see e.g. Rodríguez-Antón 2017; and Fig. 6.2). The database with the 64 Augustan towns is given in Table 6.1. Figure 6.3a presents the comparison of the full dataset (181 towns), in dark grey, and what we would expect (light grey) if the data were supposed to follow the sunrise on any day of the year homogeneously. From this figure we should expect

Fig. 6.2 A map of the western part of the Roman Empire showing the data base of the 181 Roman towns that have been measured so far. The red and pink rectangles indicate the 64 Augustan town. Red rectangles are measurements taken on-site by our team, and pink rectangles indicate measurements taken via Digital Terrain Models and orthoimages

concentrations close to the solstitial lines (vertical solid lines) and we do find such clustering, but we also find that there are other clusters within these limits that are not expected. However we cannot ascertain whether this is just a matter of how the sample was constructed or if these concentrations do actually indicate an intentionality to choose particular dates in the course of the year.

Figure 6.3b presents the results for the 64 Augustan towns in our sample (dark grey compared with the total sample so far in light grey). It is clear that now the maxima become narrower, sharper and therefore more clearly defined. It must be

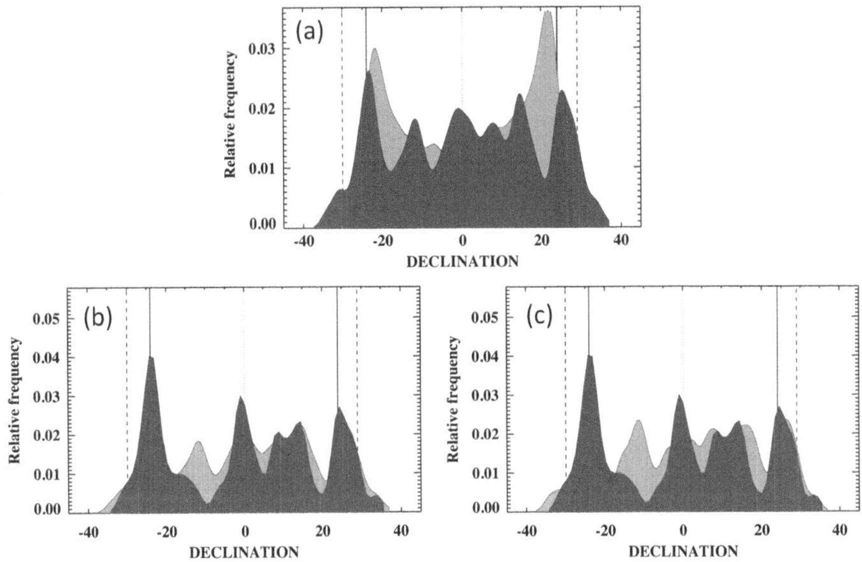

Fig. 6.3 (**a**) A declination histogram of the orientation of 181 Roman towns across the Roman west (dark grey) compared with the expected distribution of orientations in the course of a year if the orientations are supposed to follow the Sun on any day of the year (light grey). The solid vertical lines indicate the limits of the solar positions during the year (solstices), while the dashed vertical lines indicate the limits of the Moon. (**b**) Comparison of the declination histogram of the 64 Augustan towns measured in our sample and included in Table 6.1 (dark grey) with all the Roman towns measured so far by our group (light grey). (**c**) Comparison of the declination histogram of the 64 Augustan towns measured in our sample and included in Table 6.1 (dark grey) with the non-Augustan towns (light grey); vertical lines are as in Fig. 6.3a

stressed that these 64 towns are included in the total sample. Figure 6.3c makes the same comparison with the data not including the Augustan towns. From these two figures we can see that Augustan towns present a more concentrated distribution especially towards Winter Solstice sunrise (declination −24°), equinox (declination 0°) and Summer Solstice sunrise (declination 24°). There are other less well-defined concentrations near values of 10°–15°.

All these values and comparisons were done for the eastern area of the horizon. Figure 6.4 compares the orientation of the eastern and western ends of the *decumanus*. It must be noted that if the eastern end of a given decumanus is oriented towards Winter Solstice sunrise the opposite direction, towards the west will, in general point to Summer Solstice sunset. However, this is not always the case, as the actual horizon may prevent such symmetry. In the case of Fig. 6.4 if the concond were real we should expect both histograms (light and dark grey) to be mirror images of each other, and that the dark grey peak at declination −24° should have a light grey counter part at 24° of similar amplitude. But this is not the case due to the breaking of such symmetry caused by the local horizons at each town. We may observe that the concentrations are generally narrower and sharper for the eastern end, perhaps

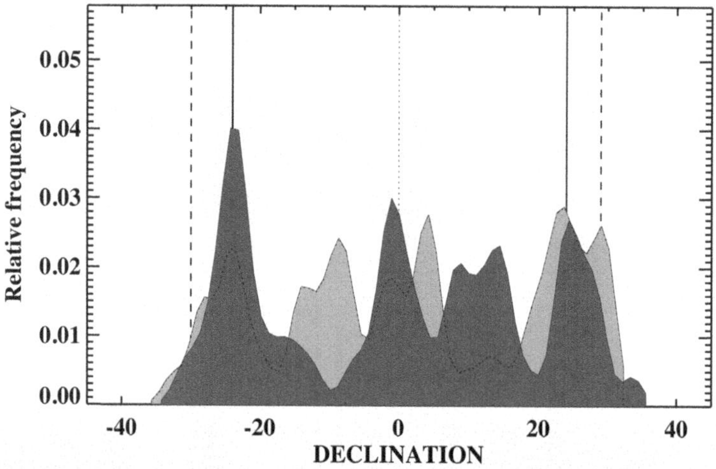

Fig. 6.4 Comparison between the eastern end of the decumani (dark grey) and the western part (light gray); vertical lines are as in Fig. 6.3a

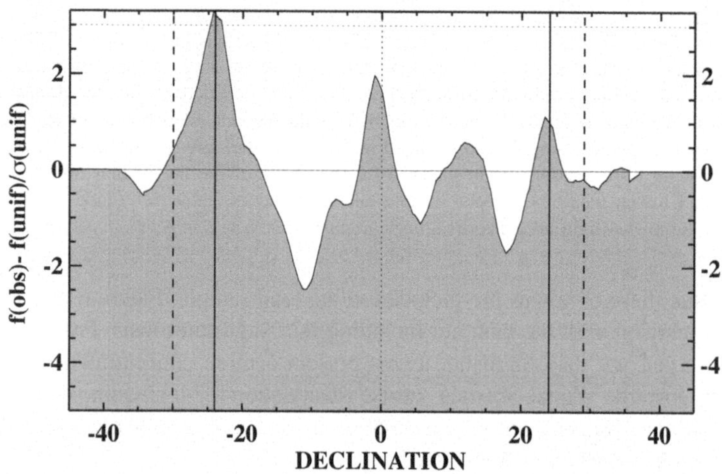

Fig. 6.5 Relative frequency of the Augustan towns (f(obs)) relative to the non-Augustan ones (f (unif)) in terms of the internal variation of the last ones; vertical lines as in Fig. 6.3a; for details see the text

indicating a better orientation in that directions, and therefore, following the scheme proposed by González-García and Sprajc (2016), an intentionallity towards that direction that fits well with the well-known prevalence of east as main cardinal point in Rome and, in general, in the whole Indo-European tradition.

Figure 6.5 makes this comparison more explicit. It directly compares the prominence of the concentrations of the Augustan towns with those in the non-Augustan ones. In this sense, this diagram provides an indication of how significant these

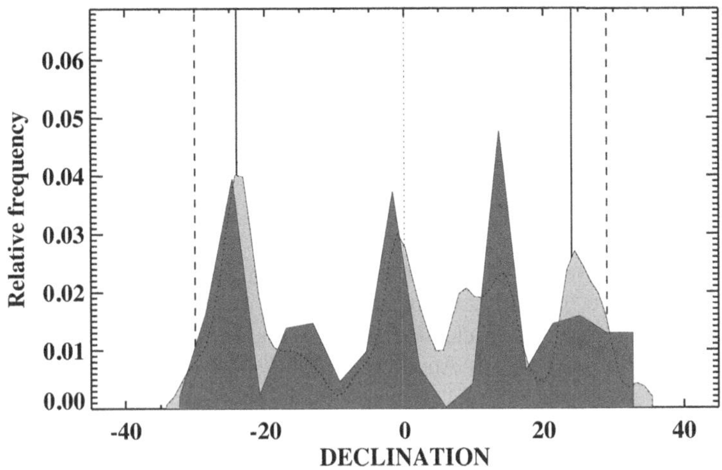

Fig. 6.6 A comparison of cities in our sample named '*Augusta*' in dark grey with all Augustan towns (light grey); vertical lines as in Fig. 6.3a

maxima are. In the natural sciences a value of 3 or above means a high relevance of the concentration, while in the humanities a value of 2 is more than acceptable. In this case, we have two clear maxima above the 2-sigma line: the Winter Solstice one and the equinoctial concentration. This clearly indicates that such directions, and possibly the dates associated with them in the Roman calendar, were of particular significance.

Finally, Fig. 6.6 includes only those towns in our sample that bear the name *Augusta*, to see if there is any difference with the rest of the Augustan settlements. We have only 18 such towns in our sample, and despite the low number statistics, we do not see any difference between this sample and the rest of towns for the three main maxima. The only major difference appears to be the concentration around declination 14°, which would correspond with the beginning of May or mid-August in a proleptic Gregorian calendar.

Discussion and Conclusions

Given the results expressed above regarding Figs. 6.1 to 6.6 we may conclude that Augustan towns were oriented following a definite pattern. Sunrise at a particular time of the year seems to have been important. For instance, Winter Solstice sunrise was a favorite choice in the orientation of Augustan towns. Other favoured times were the equinoxes and Summer Solstice. All of these were defined by a solar model that perhaps related Augustus to the Sun and, more specifically, to Apollo-Sol (Lange 2009; Rehak 2006).

In this respect, a mythical tale attributes Augustus' conception to the relationship between Apollo (transformed into a snake) and his mother *Atia* while she slept inside his temple in *Campus Martius* (Suet. *Aug*. 94, 4; the fatherhood of Augustus is also attributed to Apollo by Cassius Dio 45, 1, 2). According to Suetonius, while pregnant, *Atia* also dreamed that her innards were transported to the stars and scattered over the surface of the land and sea, and her husband Octavius dreamed that the glowing Sun emerged from Atia's womb (Suet. *Aug*. 94, 4; similarly, see Cass. Dio 45, 1, 3; Epigr. Bob. 39).

Augustus would not have hesitated to politically exploit his presumed divine filiation with Apollo from an early age. As reported by Suetonius, between 40 and 37 BCE, Octavian would have attended what was known as the 'dinner of the twelve gods' (δωδεκάθεος) dressed as Apollo (*pro Apolline ornatum*), although this presumptuous behavior was not well accepted by the populace (Suet. *Aug*. 70, 1–2; Kleiner 1988; Lange 2009; Del Hoyo 2011).

At about that time, Octavian would have promised to build a temple dedicated to Apollo during his campaign against Sextus Pompey (Suet. *Aug*. 29, 1; 29, 3; *Mon. Anc*. 19, 1; Plin. *HN* 36, 13; Vell. Pat. 2, 81, 3; Cass. Dio 49, 15, 5; Prop. 2, 31.). The construction work, which was not completed until 28 BCE, began immediately on land that he owned on the Palatine Hill alongside Augustus' house, which as stated in Suetonius had been chosen by Apollo with a lightning strike on that spot (Suet. *Aug*. 29, 3). According to the sources, a statue of Augustus stood within this area, in the guise of Apollo (Scholiast on Hor. *Epist*. 1, 3 17; Serv. Ecl. 4, 10; Kleiner 1988; Lange 2009; Rehak 2006).

Moreover, the House of Augustus had been honored through the planting of laurel trees (the sacred tree to Apollo) on both sides of the door, as well as placing the *corona ciuica* upon it (*Mon. Anc*. 34; Ov. *Fast*. 4, 951–954; *Met*. 1, 557–566). These laurel trees, in the opinion of Zanker (1988), would have given the entrance of the house a 'sacred aura'. According to Kleiner (1988: 350), "… whether Augustus wished to present himself as Apollo or simply as the god's son, he was following a venerable Eastern tradition in which a ruler was treated as a divinity." In fact, Augustus was honored in a number of Eastern towns, such as Athens, by dedicating sculptures to Neo-Apollo (allegedly dressed like this god) (Hoff 1992; Peppas-Delmousou 1979; Schmalz 2009). Furthermore, the *dies natalis Augusti* was celebrated in Athens with the typical offerings for Apollo during his feast day (*SEG* XVII, 34). Also, this connection is reflected in the sacerdotal organization in Athens at the beginning of the Principate where the priest of Rome and Augustus at the Acropolis (IG II2, 3173), Pamenes of Marathon, was also the priest of Delian Apollo (Inscr. Delos 1592; 1593; 1594; 1605; 1625; 2515; 2517 y 2518; for further discussion see Lozano Gómez 2007). Finally, a passage from Virgil (*Aen*. 8, 720–723) shows Augustus (after having been victorious in *Actium*) as ruler of the oecumene, enthroned at the blazing threshold of the Temple of Apollo, settling the matters of all countries and receiving *dona*. But the association between Augustus and the Sun is also documented for the West, since Horace (*Carm*. 4, 5, 5–8) describes Augustus' return to Rome after his stays in *Hispania*, *Gallia* and *Germania* between 16 and 13 BCE as the Sun returning to the homeland.

This question, as it was proposed in a previous work (Espinosa-Espinosa and González-García 2017), is strongly tied to the orientation of Roman towns towards sunrise during the autumn equinox. If our interpretation is correct, the idea of linking the celebration of the *dies natalis Augusti* on 23 September with this astronomical event fits more closely with the idea of Augustus' association with Apollo-Sol and, therefore, with the alignment of Roman towns towards this date. This was something that Augustus himself may well have been seeking, after being proclaimed Pontifex Maximus, by decreeing the reformation of the calendar to bring it back in line with the seasons (González-Garcia and Belmonte 2006). In fact, as has been supported by Bilić (2012), Apollo's relation with the annual solar movement in the ancient world seems clearly based on ancient Greek narrative, with an emphasis on the solstices. This identification appears to have continued in Roman times, as illustrated by Cicero (*N.D.* 3, 57) and Claudian (*De VI Cons. Hon.* 26–27). In fact, Herodian (5, 6, 6–8) describes a summer festival performed by Elagabalus in Rome, in which the Emperor ran backwards in front of a solar chariot driven by the deity. His running backwards, in the opinion of Bilić (2012), certainly symbolized the start of the Sun's backwards movement towards the south following the Summer Solstice.

A different case seems to be at stake with respect to the possibility of the fourth concentration related to the beginning of May or mid August. There is no apparent Roman festival that can be clearly related with this concentration of orientations. Given the geographic location of the towns producing this concentration, this could be related to the culture previously existing in such areas. In this case, it is remarkable that a good number of those towns (5 out of 8) are from the Iberian Northwest where the classical writers indicate that the tribes had a Celtic cultural background, also attested by the theonims and toponims with Celtic roots found in these regions (González García 2011; Luján Martínez 2009). Thus, in such areas there could have been a process of explicit or implicit negotiation between the need to impose the Romanness of the towns while including Celtic orientations, something similar to what has been mentioned earlier for *Lugdunum*, for the Gaulic tribes, or for *Carthago Nova* for the Punic population.

If we must suppose those towns to be the result of negotiations between the Romans trying to impose their model and the local elites trying to become Roman, it is interesting to see how such a model was implemented in different cases and how compromise was reflected in the orientation of towns. In certain areas this would imply a full acceptance of the Roman symbols, including in them the use of the Roman orientations, meaning those towards Winter Solstice and perhaps the Equinox, as directly related with Augustus. In other cases instead, such negotiation meant a mixture of a Roman town with local orientations. It seems remarkable in this sense that such cases do seem to point to the importance and resilience of Celtic elements. However, this last point could be the result of a certain selection effect, as our sample concentrates largely in areas where the Celtic population could have been important. It seems clear that in order to elucidate this issue we need further dedicated research in these areas, together with an investigation on what happened in other regions of the Roman Empire at the same time.

In addition, it would be interesting to see what happened before Augustus. Julius Caesar for instance founded a large number of towns as well. We could mention two that might seem indicative: Carthage and Corinth. Both cities, destroyed at 146 BC were refounded as Roman colonies at the time of Caesar, and it is interesting to note that *Carthago* was oriented parallel to the coast line, but in such a way that is at the same time oriented towards Winter Solstice sunrise (Esteban 2003). Meanwhile, the Roman layout of Corith has a skewd orientation of 3° with respect to the cardinal points (Gilman Romano 2003: 285). Such orientation, and taking into account the height of the horizon, indicates an orientation towards sunrise on 25 March, in the proleptic Gregorian Calendar. It is interesting that such a date was the traditional date for the equinox and was probably one of the dates that Julius Caesar considered when reforming the Republican Calendar to bring it in step with the seasons (González-Garcia and Belmonte 2006).

It is highly indicative that both orientations appear as the most frequent ones in the latter Augustan foundings. In this sense it would also be advisable to extend this research in order to see if the settlements founded by Pompeius Magnus in the East do have a systematic orientation like those of later Roman rulers, to try and resolve whether the Hellenistic tradition came with him from the East.

References

Barton, T. (1995). Augustus and Capricorn: Astrological polyvalency and imperial rhetoric. *The Journal of Roman Studies, 85*, 33–51.

Belmonte, J. A., & González-García, A. C. (2010). Antiocho's Hierathesion at Nemrud Dag revisited: Adjusting the date in the light of astronomical evidence. *Journal for the History of Astronomy, 41*, 469–482.

Belmonte, J. A., Noguera, J. M., González-García, A. C., & Rodríguez-Antón, A. (2016). The materialization of sacred space: Astronomy and landscape in Carthago Nova. In T. Lonsladen & F. Ventura (Eds.), *The materiality on the sky, Proceedings of the SEAC 2014 Conference* (pp. 65–78). Bath: Sophia Centre Press.

Bertarione, S. V., & Magli, G. (2015). Augustus' power from the stars and the foundation of Augusta Praetoria Salassorum. *Cambridge Archaeological Journal, 25*(1), 1–15.

Bilić, T. (2012). Apollo, Helios, and the solstices in the Athenian, Delphian, and Delian calendars. *Numen, 59*, 509–532.

Briquel, D. (2008). L'espace consacré chez les Étrusques: Reflexions sur le rituel étrusco-romain de fondation des cites. In X. Dupré Raventós, S. Ribichini, & S. Verger (Eds.), *Saturnia Tellus. Definizioni dello Spazio Consacrato in Ambiente Etruco, Italico, Fenicio-Punico, Iberico e Céltico* (pp. 27–47). Rome: Consiglio Nazionale delle Ricerche.

Del Hoyo, J. (2011). Aprovechamiento político de los dioses por Augusto y su tiempo. In M. Campo (Ed.), *Mites, Ofrenes Funeràries i Monedes. XV Curs d'Història Monetària d'Hispània. Barcelona, 24 i 25 de Novembre de 2011* (pp. 45–53). Barcelona: Museu Nacional d'Art de Catalunya.

Espinosa-Espinosa, D., & González-García, A. C. (2017). A.D. VIIII Kalendas Octobres, Dies Natalis Augusti. Some considerations on the astronomical orientation of Roman Cologne and the imperial cult. *Numen, 64*, 1–23.

Esteban, C. (2003). Temples and astronomy in Carthage. In *Calendars, symbols, and orientations: Legacies of astronomy in culture* (pp. 135–142, (Observatory Reports, 59). Uppsala: Uppsala Astronomy.

Ferro, L., & Magli, G. (2012). The astronomical orientation of the urban plan of Alexandria. *Oxford Journal of Archaeology, 31*, 381–389.

Galinsky, K. (2012). *Augustus: Introduction to the life of an emperor*. Cambridge: Cambridge University Press.

García Quintela, M. V., & González-García, A. C. (2014). Le 1er Août à *Lugdunum* sous l'Empire Romain: Bilans et nouvelles perspectives. *Revue Archáeologique de l'Est, 63*, 157–177.

Gilman Romano, D. (2003). City planning, centuriation and land division in Roman Corinth: Colonia Laus Iulia Corinthiensis & Colonia Iulia Falvia Augusta Corinthiensis. In *Corinth 20, The Centenary: 1896–1996* (pp. 279–301). Athens: The American School of Classical Studies.

González, R., & Ruiz de Arbulo, J. (Eds.). (2010). *Simulacra Romae II. Rome, les Capitals de Province (Capita Provinciarum) et la Création d'un Espace Commun Européen. Une Aproche Archéologique*. Reims: Societé Archéologique Champenoise.

González García, F. J. (2011). Los Célticos de Gallaecia: Apuntes sobre etnicidad y territorialidad en la Edad del Hierro del Noroeste de la Península Ibérica. *Complutum, 22*(1), 117–132.

González-Garcia, A. C., & Belmonte, J. A. (2006). Which equinox? *Archaeoastronomy: Journal for Astronomy in Culture, 20*, 95–105.

González-García, A. C., & García Quintela, M. V. (2014). The 1st of August at Lugdunum: Astronomy and imperial cult in Galia. *Mediterranean Archaeology and Archaeometry, 14*(3), 83–91.

González-García, A. C., & Magli, G. (2014). Roman city planning and spatial organization. In C. L. N. Ruggles (Ed.), *Handbook of archaeoastronomy and ethnoastronomy* (pp. 1643–1650). Heidelberg: Springer.

González-García, A. C., & Sprajc, I. (2016). Astronomical significance of architectural orientations in the Maya Lowlands: A statistical approach. *Journal of Archaeological Science: Reports, 9*, 191–202.

González-García, A. C., Rodríguez-Antón, A., & Belmonte, J. A. (2014). The orientation of Roman towns in Hispania: Preliminary results. *Mediterranean Archaeology and Archaeometry, 14*(3), 107–119.

González-García, A. C., et al. (2015). Orientatio ad Sidera: Astronomía y Paisaje urbano en Qart Hadašt/Carthago Nova. *Zephyrus, 75*, 141–162.

Gros, P., & Torelli, M. (1994). *Storia dell'Urbanistica. Il Mondo Romano*. Roma: Editori Laterza.

Hannah, R., & Magli, G. (2011). The role of the Sun in the Pantheon's design and meaning. *Numen, 58*, 486–513.

Haselberger, L. (2014). The Horologium of Augustus: Debate and context. *Journal of Roman Archaeology*, Supplementary Series, 99.

Hoff, M. C. (1992). Augustus, Apollo and Athens. *Museum Helveticum, 49*, 223–233.

Kaiser, A. (2011). *Roman urban street networks*. New York: Routledge.

Kleiner, F. S. (1988). The arch in honour of C. octavius and the fathers of Augustus. *Historia, 37*(3), 347–357.

Lange, C. H. (2009). *Res Publica Constituta. Actium, Apollo and the Accomplishment of the Triumviral Assignment*. Leiden-Boston: Brill.

Laurence, R., Esmonde Cleary, S., & Sears, G. (2011). *The city in the Roman West c.250 BC–c.AD 250*. Cambridge: Cambridge University Press.

Lozano Gómez, F. (2007). Los agones de los Augustos en Atenas. In M. Mayer, G. Baratta, & A. Guzmán (Eds.), *Provinciae Imperii Romani Inscriptionibus Descriptae* (pp. 851–856). Barcelona: Institut d'Estudis Catalans.

Luján Martínez, E. R. (2009). Pueblos celtas y no celtas de la Galicia antigua. Fuentes literarias frente a fuentes epigráficas. In J. Aparicio Pérez & L. Silgo Gauche (Eds.), *Real Acadèmia de Cultura Valenciana. Sección de Estudios Ibéricos D. Fletcher Valls. Estudios de Lenguas y Epigrafía Antiguas-ELEA n° 9* (pp. 219–250). Valencia: Real Acadèmia de Cultura Valenciana.

Magli, G. (2008). On the orientation of Roman towns in Italy. *Oxford Journal of Archaeology, 27*(1), 63–71.
Owens, E. J. (1992). *The city in the Greek and Roman World*. New York: Routledge.
Pankenier, D. W. (2014). Astronomy and city planning in China. In C. L. N. Ruggles (Ed.), *Handbook of archaeoastronomy and ethnoastronomy* (pp. 1643–1650). Heidelberg: Springer.
Peppas-Delmousou, D. (1979). A statue base for Augustus IG II2 3262 + IG II2 4725. *American Journal of Philology, 100*, 125–132.
Ramsey, J. T., & Licht, A. L. (1997). *The comet of 44 B.C. and Caesar's funeral games*. Atlanta, GA: Scholars Press.
Rehak, P. (2006). *Imperium and Cosmos. Augustus and the Northern Campus Martius*. Madison: University of Wisconsin Press.
Rodríguez-Antón, A. (2017). *Cosmovisión y Urbanismo en la Roma Antigua: Orientación de Ciudades y Campamentos Romanos*. Ph.D. thesis, Universidad de La Laguna, Spain.
Rodríguez-Antón, A., González-García, A. C., & Belmonte, J. A. (2017). Estimating the reliability of digital data acquisition in cultural astronomy. Its use in the case of Roman North Africa. *Journal of Skyscape Archaeology, 3*(2), 191–206.
Rykwert, J. (1988). *The idea of a town. The anthropology of urban form in Rome, Italy and the ancient world*. Princeton: Princeton University Press.
Schmalz, G. C. R. (2009). *Augustan and Julio-Claudian Athens. A new epigraphy and prosopography*. Leiden-Boston: Brill.
Woodward, P., & Woodward, A. (2004). Dedicating the town: Urban foundation deposits in Roman Britain. *World Archaeology, 36*(1), 68–86.
Zanker, P. (1988). *The power of images in the age of Augustus*. Ann Arbor, MI: University of Michigan Press.

Chapter 7
The *Uaratio* and Its Possible Use in Roman Urban Planning to Obtain Astronomical Orientations

Andrea Rodríguez-Antón, Margarita Orfila Pons, A. César González-García, and Juan Belmonte Aviles

Abstract Several works have tried either to demonstrate or reject the notion that the orientation of the main axis of a Roman city was deliberate since its choice might add an extra sacred dimension to the entire urban space [González-García et al. (Mediterranean Archaeology and Archaeometry 14(3):107–119, 2014; Magli (Oxford Journal of Archaeology 21(6):63–71, 2008)]. There exist ancient texts that support the hypothesis of the existence of astronomical orientations, such as those of Frontinus (*De Agrimmensura*, 27) or Hyginus Gromaticus (*Constitutio*, I). In the case that these precepts were fulfilled: how to achieve it? Besides the astronomical hypothesis, some scholars have pointed to the use of a geometrical technique: the *uaratio* (Orfila et al. La orientación de las estructuras ortogonales de nueva planta en época romana. De la varatio y sus variaciones. 2014). By this, the short sides of a regular triangle that are in ratios of integer numbers (for example 1:2, 2:3) are laid along the cardinal axes. In this work we present a comparison of the orientation of 81 Roman towns in the Iberian Peninsula, measured in situ, with *uaratio* angles with aspect ratios up to 12:12. By this exercise we want to discern whether the orientations were astronomical, purely geometrical, or if geometry could have fostered astronomical aims by using selected and well-known angles to trace lines that fitted the desired astronomical purposes. It is then, an attempt to shed more light to the issue of the orientation of Roman towns by combining two hypotheses that, in contrast to what it might seem, could be complementary but not contrary.

A. Rodríguez-Antón (✉) · J. B. Aviles
Instituto de Astrofísica de Canarias & Universidad de La Laguna, Santa Cruz de Tenerife, Spain
e-mail: ara@iac.es; jba@iac.es

M. Orfila Pons
Universidad de Granada, Granada, Spain
e-mail: orfila@ugr.es

A. C. González-García
Instituto de Ciencias del Patrimonio, Incipit-CSIC, Santiago de Compostela, Spain
e-mail: a.cesar.gonzalez-garcia@incipit.csic.es

G. Magli et al. (eds.), *Archaeoastronomy in the Roman World*, Historical & Cultural Astronomy, https://doi.org/10.1007/978-3-319-97007-3_7

Introduction

Roman society is characterised, among other factors, by the incorporation of norms used in daily life in measuring units, in time reckoning and in modular division when constructing buildings. These norms also were applied during the division and careful planning of conquered or confiscated territory.

At that time, land division played a key role in the expansionist process and was essential in reorganizing the territory after a conquest, when new structures, such as towns, were erected. The foundation or re-foundation of towns involved important decisions about their spatial organization and orientation (Orfila Pons et al. 2017a). Furthermore, Romans accepted the idea of well-planned cities, and urban organization was applied with common standards and objectives. Importantly, it provided a reliable means of calculating the taxes that should be collected per each plot of land (Gilman Romano 2003).

Land division was carried out by a number of individuals: the *curator operis*, who coordinated the project, the *architectus* or *mensor* who defined the plan, and the *gromatici* or *agrimensores*, who laid out the divisions. They also measured irregular plots, understood the complexities of land law, and by the Late Empire even became judges in land disputes (Lewis 2001). Each plot of land, or *centuria*, was demarcated by *limites* or roads at right angles known as *decumani* and *cardine* that, ideally, should run east-west and north-south, respectively. However, previous studies have revealed that cardinal orientation was not a common trend (González-García and Magli 2015; González-García et al. 2014; Orfila et al. 2014).

Besides the practical component, a symbolic aspect should not be dismissed, since the creation or a new *territorium* contained an ideological factor (Castillo Pascual 1996). The foundation of new urban entities was accompanied by the performance of a ritual that might have influenced the selection of the spatial disposition of the main axes (Rykwert 1988). In this sense, by considering a combination of practicality, adaptation to the terrain conditions and symbolism, an unavoidable question arises about the motivations that were used in selecting the orientation of Roman urban grids.

Although some of the first publications on this topic (Le Gall 1975; Peterson 2007) refused to accept the existence of any intentionality behind the orientation of Roman towns, a number of later studies suggested the presence of well-defined patterns of orientation in different regions across the Roman Empire (Orfila et al. 2014; Richardson 2005). Furthermore, astronomy-based explanations for the orientation of many of the settlements in the studied samples were proposed by a number of authors (González-García and Costa Ferrer 2011; González-García et al. 2014, 2015). If this was so, how did Roman surveyors achieve these astronomical designs?

In this work we propose a geometrical method that Roman topographers might have applied to urban layout in order to set the direction of the streets according to some celestial events. It consists of the application of a geometrical technique used in Roman times and explained by the ancient writer Nypsius: the *uaratio*. Basically, this technique is based on the use of right-angled triangles where the length of its short sides are in ratios of integer numbers or where the lengths of the three sides are in ratios of integer numbers. These last are called Pythagorean triangles because their sides form a Pythagorean triple (like 3:4:5 for instance, that obey the Pythagoras Theorem).

The possible use of a standardized geometrical method to determine the orientation of Roman towns has been explored previously (Orfila 2012; Orfila and Moranta 2001; Orfila et al. 2014; Orfila Pons et al. 2017b). In order to test this hypothesis, we compared the orientations of 76 Roman settlements in the Iberian Peninsula, which were measured in situ, with a set of *uaratio* angles. This test was complemented by the results of practical experiments with reconstructions of instrumentation used by the *agrimensores*, such as the *groma, metae* and a *gnomon*, developed by the SOTOER group at Universidad de Granada (Costa and Orfila 2014; Orfila et al. 2014).

If the astronomical and the geometrical hypotheses converge, we might have found a relatively simple technique that worked as a standard procedure in the various processes of expansion (Adams 1982).

The *Uaratio*: A Proposal for Its Implementation

The *uaratio* was a technique used by Roman topographers, and was described by Marcus Iunius Nypsius in the second century CE (*Fluminis uaratio*; *Limitis Repositio*: La. 285.1–295.15, Bouma 1993). Its application allowed the identification of the *limites* of the land lots assigned to each community. It was also useful in determining the orientation of every orthogonal project when the land lots were not cardinally oriented (Orfila Pons et al. 2017b; Roth-Congés 1996). These interpretations are included in the *Corpus* of the *Gromatici Veteres*, which Chouquer (2004) interpreted and associated with the cadastre (land registry) that was made after the fire at the Roman *archivum* in CE 80, under Vespasian. In this, the location and limits of the different pieces of land was recorded.

On the basis that every architectural endeavour must start from a primordial line, we assume that this line was the meridian, as proposed by some scholars (see e.g. Orfila and Moranta 2001; Orfila et al. 2014; Orfila Pons et al. 2017b). The meridian line does not depend on the site location, and it is relatively easy to obtain by using a gnomon. This procedure is described by ancient authors such as Vitruvius (*De Architectura* I, 6.6) and Hyginus Gromaticus (*Constitutio* I, 52.4–22). As mentioned above, in most cases the urban axes were not cardinally oriented but were at different angles to the meridian line. Our proposal is that this may have originated from the use of the *uaratio*, as described below.

We follow Orfila et al. (2014) and consider two possible ways that this technique could have been applied. The first one consists of tracing the sides of a right-angled triangle, whose lengths present integer ratios, on the cardinal axes previosly obtained with a gnomon, as shown in Fig. 7.1(3ab). The ratios are called *uariationes*. The hypotenuse of the resulting triangle defines the direction of one of the streets, and

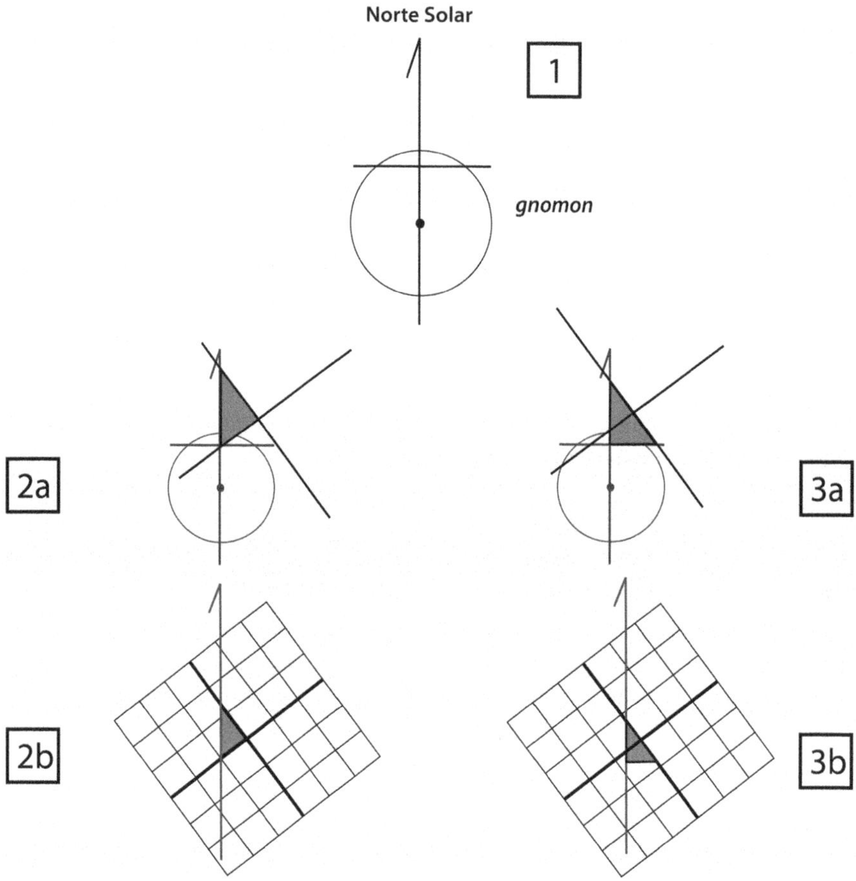

Fig. 7.1 The two hypothetical procedures for the application of the *uaratio* to obtain the direction of a new urban grid (adapted from Orfila et al. 2014)

that would depend on the ratio chosen for the triangle sides (1:2, 2:3, 3:5, etc). The perpendicular street could have been obtained by using a *groma*.[1]

Alternatively, this technique can be used by considering right-angled triangles whose sides are Pythagorean triples.[2] Our proposal is that the hypotenuse was drawn along the meridian line and the sides of the triangle, both with lengths that used integer numbers, would define the direction of the urban grid as seen in Fig. 7.1(2ab) (see Orfila et al. 2014). This idea was first proposed by Margarita Orfila and Luis Moranta (2001) for the lowest Pythagorean triples of 3:4:5 and 5:12:13. In addition,

[1]This is an instrument based on a horizontal cross with arms in right angles, on a vertical support. From the end of each of the four arms hung a cord or plum-line tensioned by a bob. It was used for sighting and setting out straight lines and right angles (Lewis 2001).

[2]These have three positive integers (a, b, c) such that $a^2 + b^2 = c^2$.

the fact that the use of fractions to obtain irrational numbers was common in Antiquity (Neugebauer 1969; Peterson 1992) constitutes further motivation to explore their possible application in determining astronomical orientations of *centuriae* and urban grids.

Sample and Methodology

The first step was to test whether the *uaratio* was present in the city plans. To accomplish this, we compared the azimuth of the orientation of a number of Roman cities in the Iberian Peninsula with the distribution of *uaratio* angles.

The sample contains the azimuths of the main streets of 76 Roman towns in ancient *Hispania*, corresponding to 81 orientations (Fig. 7.2) (Rodríguez-Antón 2017). This is due to the presence of more than one urban grid in some sites, such as *Italica* or *Corduba*. The data were collected in situ during several fieldwork campaigns (Rodríguez-Antón 2017). The altitude of the horizon in the direction marked by each city grid was also measured in order to compute the astronomical declination of a hypothetical object that would rise or set at that specific point on the horizon.

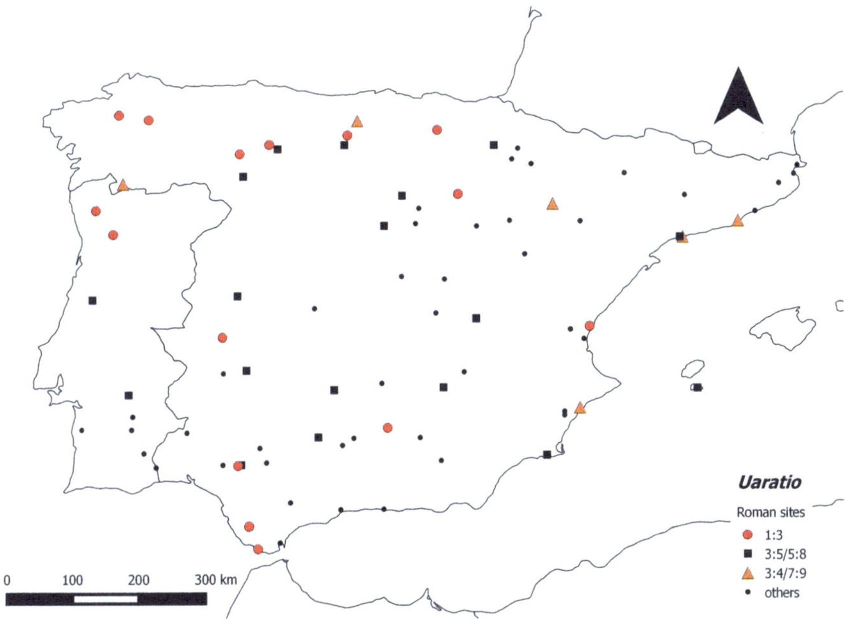

Fig. 7.2 Map showing the Roman cities studied, classified by the azimuth groups present in the sample that fit low *uariationes*

We used a professional compass to obtain the azimuths and a clinometer to measure the altitudes of the horizon. These instruments introduce nominal errors of ±¼° and ±½° respectively. However, azimuth errors were sometimes larger owing to external factors, such as the preservation state of the structures under consideration. The measured azimuths were corrected for magnetic declination, which was either estimated from the model WMM2012, or updated ones, available at: https://www.ngdc.noaa.gov/geomag-web/#declination. When altitudes could not be measured due to blocked horizons, we used reconstructions of the horizon from the Digital Terrain Model SRTM (Shuttle Radar Topography Mission) of NASA through the on-line panorama generator HeywhatsThat, available at: http://www.heywhatsthat.com/.

The comparison sample is a dataset of angles derived from the application of the *uaratio* as explained in the previous section. We considered *uariationes* ranging from 1:1 to 12:12. Since equivalent fractions correspond to equal angles, these were considered only once. That is, we did not include the angles resulting from, e.g., ratios 1:3, 2:6 and 3:9, but just those of 1:3 (Fig. 7.3). Figure 7.4a shows the computed distribution of the *uariationes* computed from 1:1 to 12:12 across the entire horizon.

A density distribution with an Epanechnikov kernel has been used to calculate the histograms in the present work. The bandwidth in each case depends on the sample size and the errors in the measurements, so that fine details can be appreciated while avoiding the over-smoothing of the distribution (González-García and Šprajc 2016). The value of the bandwidth is indicated in each particular case.

In the next section we will compare *uaratio* angles with azimuths (Fig. 7.4b), and angles bearing from north in the sample from *Hispania* with possible *uariationes* and angles from triangles in relation of Pythagorean triples (Fig. 7.5). Finally, we will compute the declinations of the orientations of the sample from *Hispania* to test whether the main values fit the use of simple *uariationes* (Fig. 7.6).

Testing the *Uaratio* on Roman Settlements

Due to the orthogonality of the street grids, the azimuth distribution, as well as the *uaratio* angles, are repeated four times along the 360° of the horizon (Fig. 7.4a). This allowed us to limit the analysis to a 90° wide sector in order to better appreciate the coincidence or divergences of both angle distributions (Fig. 7.4b). Due to the azimuth error, in general estimated in ±0.5°, and the sample size (81 measurements) the bandwidth in the histograms in Fig. 7.4 is 1°. We established a margin of ±2° to consider that a ratio of *uaratio* fitted a city orientation. For that we took into account the intrinsic error of the azimuths and the irregularities that could have derived during the tracing of the meridian line and the triangles on the ground.

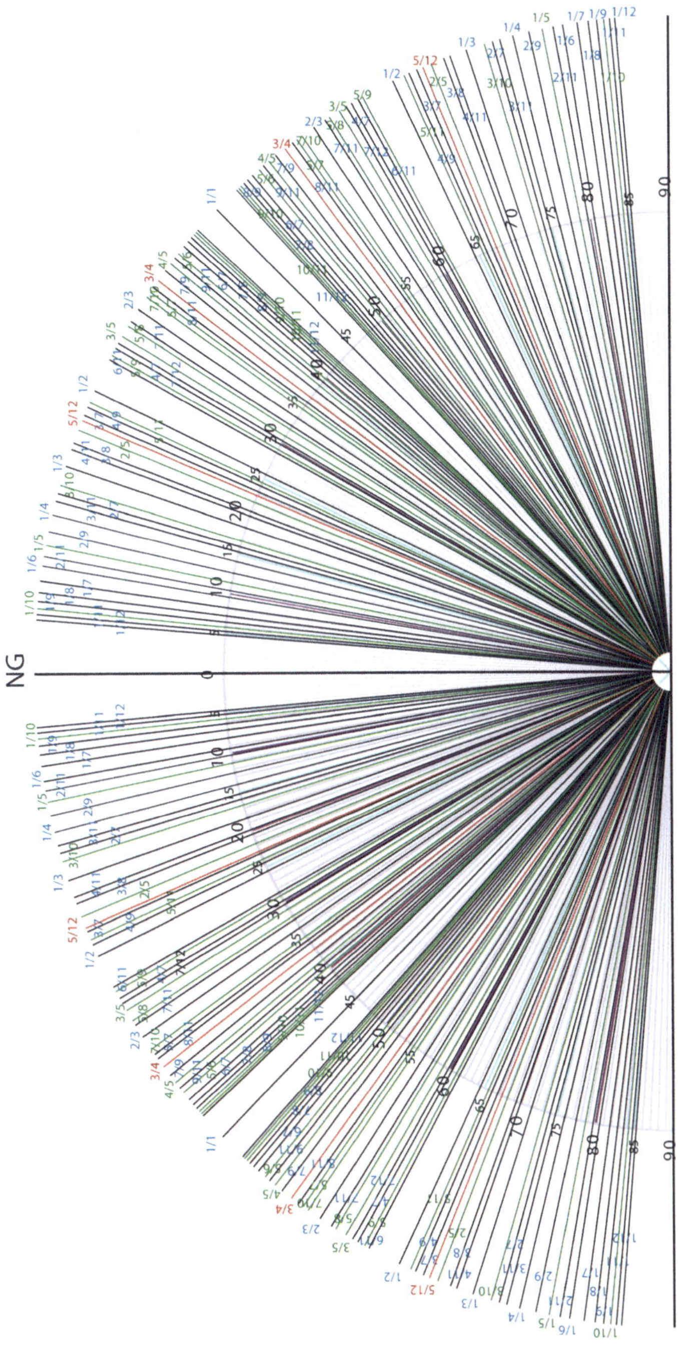

Fig. 7.3 Possible *uariationes* by using fractions from 1:1 to 12:12. 47 orientations per sector of 45°. Pythagorean uariationes in red (adapted from Orfila et al. 2014)

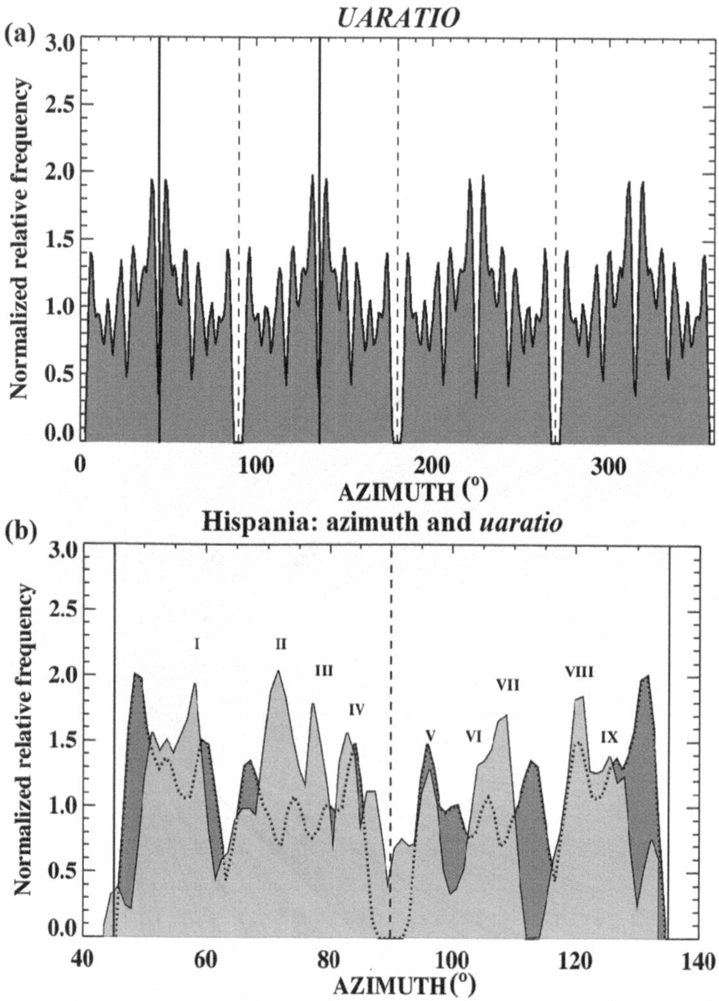

Fig. 7.4 (**a**) Distribution of *uaratio* angles computed from *uariationes* 1:1 to 12:12. Equivalent proportions are considered just once. Dashed vertical lines indicate the cardinal points and solid vertical lines the interval of 90° wide [45°, 135°] chosen for (**b**). (**b**) Distribution of *uaratio* angles from the interval indicated in (**a**) within vertical solid lines (dark grey) and the azimuths of the *decumani* of the sample of Roman towns in *Hispania* (light grey). Both distributions are the same in the three remaining symmetrical divisions. Roman numbers indicate the azimuth groups of the sample, whose equivalent values are shown in Table 7.1; read the text for further details

The election of the azimuth sector [45°, 135°] is because we define a decumanus—the street that should run east-west—within that range and within its symmetrical interval towards the west [225°, 315°]. At latitudes of the Iberian Peninsula, both intervals comprise the azimuth sectors within which the Sun and the Moon move during a full solar or lunar cycle, so it is adequate to compare the azimuths of the *decumani* with solar positions. It is interesting to note that the azimuth distribution in Fig. 7.4b presents well-defined orientation groups, and that

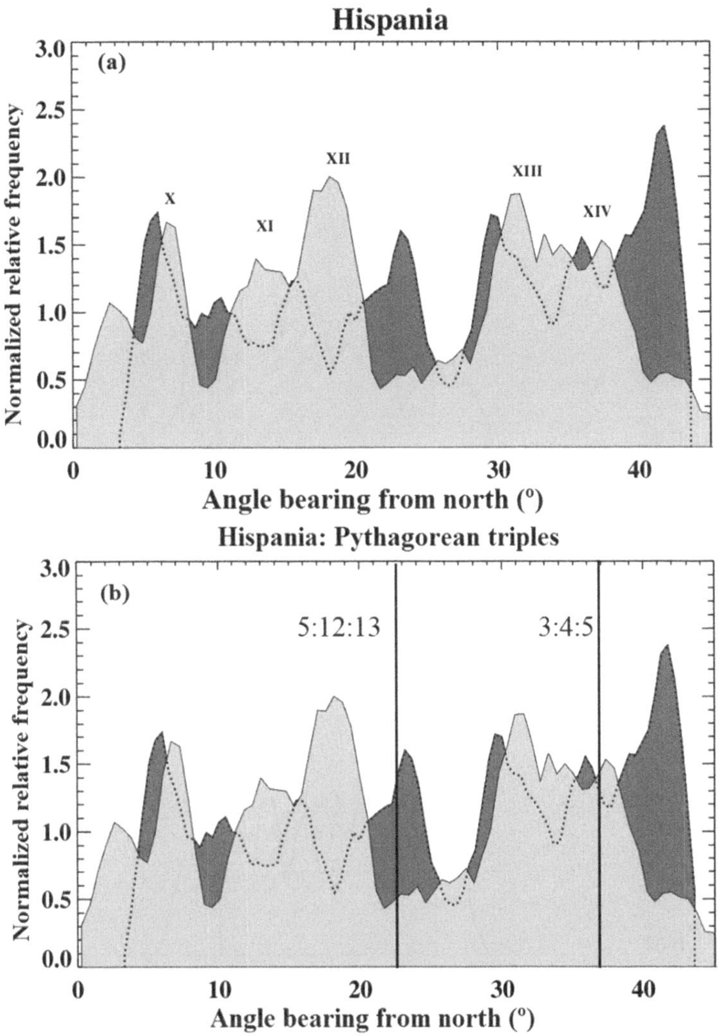

Fig. 7.5 Histograms with the orientation of the angles bearing from north of the cities in the sample in Hispania (light grey) and *uaratio* angles within an interval of [0°, 45°] (dark grey). Roman numbers in (**a**) are described in Table 7.1. Solid vertical lines in (**b**) indicate the equivalent angles by using Pythagorean triples to orientate towns

the main minima of the distributions of both datasets coincide. This means that there is a lack of cities oriented towards directions whose angles do not correspond to a *uaratio* proportion.

In order to further simplify the comparison, we took into account the symmetry of the *uaratio* angles within a 90° sector. That is, we considered symmetrical proportions together (e.g., 1:3 and 3:1 are considered to be equal). Since in this study

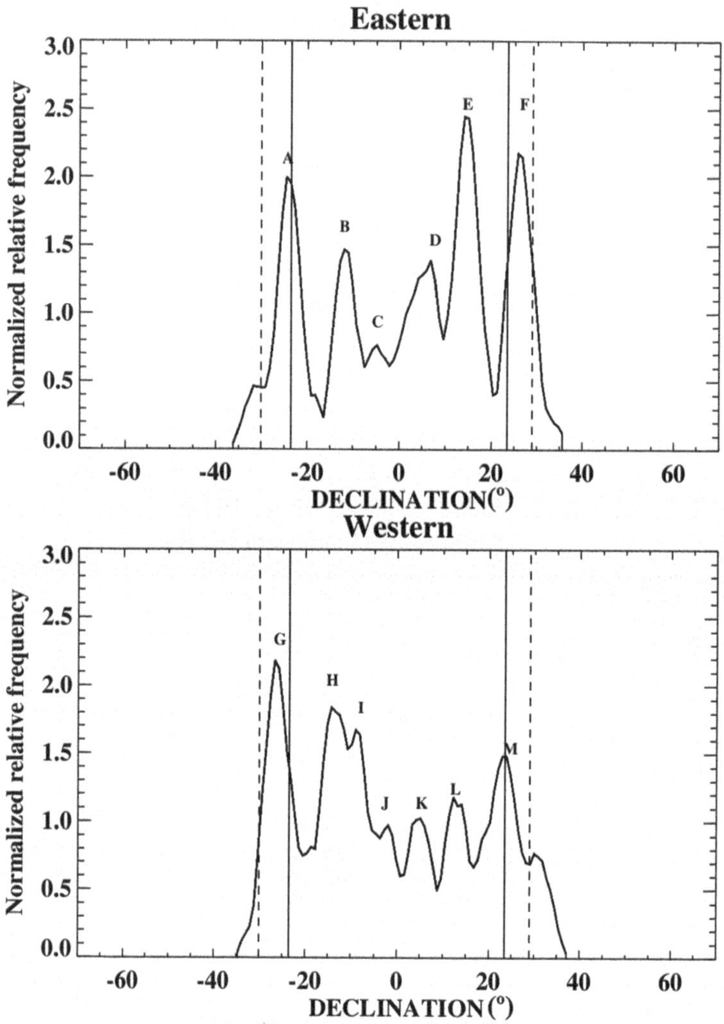

Fig. 7.6 Declination histograms of the *decumani* of the sample of Roman towns in Hispania, defined within azimuth sectors of [45°, 135°] towards the east and [225°, 315°] towards the west. Vertical solid lines represent the declination of the Sun at the solstices and dashed lines those of the Moon at its major lunastices

our main reference is the local meridian, we focused on the range [0°, 45°]. To take advantage of the mentioned symmetry, we converted the azimuths in the sector [45°, 90°] to the range [0°, 45°] by considering their value relating to 45° (called Φ) and using the formula $\alpha = 45° - \Phi$. Now α is the symmetric angle to Φ bearing from 45° (Fig. 7.5). Angles in the range of [0°, 45°] were left unmodified. We stress that with this choice in our analysis now concerns the application of particular proportions rather than the orientation of the town. The angles of the maxima of Fig. 7.5a and the

Table 7.1 Values for angles of the maxima in the histogram in Fig. 7.5a, *uariationes* that fit these angles, and in the last column equivalent peaks from Fig. 7.4b

Maxima	A (°)	*Uaratio* (°)	X:Y	Equivalent maxima
X	7	63	1:9	IV
		8.1	1:7	V
		7.1	1:8	
XI	12	11.3	1:5	III
		12.5	2:9	VI
XII	18	18.4	1:3	II/VII
XIII	31.5	30.5	7:12	I
		30.9	3:5	VIII
		32.0	5:8	
		32.5	7:11	
XIV	37	36.0	8:11	IX
		36.9	3:4	
		38.5	4:5	
		37.9	7:9	

possible *uariationes* that fit each azimuth are shown in Table 7.1. What is observed in the histograms in Fig. 7.5 is that symmetrical peaks relating to the cardinal points in Fig. 7.4b merge and give those wider maxima in Fig. 7.5 (Table 7.1).

To compute the histogram we used a density distribution with an Epanechnikov kernel with a bandwidth of ¾°, since now there are more elements within a smaller interval and it is convenient in order to better appreciate fine details in the distributions. Once again, there is a coincidence of the main minima of both distributions. Furthermore, it is interesting to note that three of those maxima (XII, XIII and XIV, Table 7.1) fit angles that derive from relatively simple *uariationes* such as 1:3 or 3:4.

We next tested whether city azimuths matched the angles resulting from implementing right-angled triangles with sides in relation to Pythagorean triples 3:4:5 and 5:12:13. The reason for using these triples is that these are the lowest ones; that is, the ones that represent the smallest fractions. Additionally, these ratios were often used in Antiquity (Gros 1976; Orfila 2014; Roskams 2001). As can be shown in Fig. 7.5b, there is a quite appreciable group of azimuths that approximate to the equivalent angles of using 3:4:5, while not many match the angles resulting from 5:12:13.

Finally, we conducted a Kolmogorov-Smirnov test to broaden our understanding of the results and to check whether both datasets were drawn from the same parent population. That is, if the azimuths of these groups of Roman settlements could have been obtained by using the *uaratio*. If the resulting probability was low enough, the idea that both samples had the same origin could be excluded. A probability of 0.05 or smaller is required to dismiss this hypothesis with a confidence level of 95%. In this case the test yields a probability of 0.14, so we cannot discard the possibility that both samples come from the same parent population with sufficient confidence level.

A Possible Link Between *Uariationes* and Astronomical Positions

As seen above, the use of the *uaratio* to determine the orientation of new cities cannot be discarded. In case this was indeed a common practice, what could have motivated the election of specific angles?

Certain texts by ancient topographers such as Frontinus (*De Agrimensura*, 27) and Hyginus Gromaticus (*Consitutio* 1) contain practical recipes for land demarcation and surveys (Gilman Romano 2003). In particular, they indicate the necessity to follow the path of the Sun when orienting the decumanus of a town. The sacred character of this task and the performance of a ritual in the foundation of a new urban, or even military, space should not be underestimated either (Rykwert 1988), and give support to the astronomical hypothesis for the orientation of towns. Furthermore, a number of studies also suggest that the main streets of several Roman towns could have been laid out according to some celestial phenomena, both in the Iberian Peninsula (e.g. González-García and Costa Ferrer 2011; González-García et al. 2014, 2015) and in other regions of the Roman Empire (e.g. González-García and Magli 2015; Rodríguez-Antón et al. 2016a, b).

In this exercise we have selected the lowest *uariationes*, or fractions, from Table 7.1 whose equivalent angles fit maxima of the sample present in the diagram in Fig. 7.5 and calculated the resulting declination in the direction of the hypothetical *decumani* that bear azimuths equal to these *uaratio* angles. We considered a flat horizon ($h = 0°$) and latitudes of 40° and 42°, because these are approximately the mean values for the Iberian Peninsula and Rome respectively.

By this, our aim was to check whether astronomical orientations were obtained by the standardized application of specific *uariationes* when observations of astronomical rises and sets could not be observed directly. The idea of comparing the declinations for the latitude of Rome arises from the possibility that, in case the *uaratio* were actually used to point towards particular celestial objects, the *Urbs* might have been the reference places when the standards were stablished and later exported to different regions.

Assuming that the application of *uariationes* that involved low proportions would have been easier, and thus more common, we began this test by exploring the resulting declination values from the use of the simplest ratios that fitted some of the maxima of the azimuth in the sample. In particular, we calculated the declination of hypothetical *decumani* whose azimuths coincided with the corresponding angles of using the lowest *uariationes* in Table 7.1. From the peaks in Fig. 7.5, the orientation groups of the sample that derived from low *uaratio* proportions were XII (1:3), XIII (3:5) and XIV (3:4 or 4:5). Maxima X and XI coincide with angles that result from bigger proportions; maybe with the exception of 1:5 in XI.

It should be highlighted that because of the estimated error margin to differentiate an azimuth from a *uaratio* proportion (of ±2°), and the significant density of possible *uaratio* angles within each interval of 45° width, there is normally more than one fraction associated with each azimuth. In these cases we tend to assume that the lower the relation of the triangle sides, the more common the use of that fraction.

It must be stressed that not all the angles in Fig. 7.5 and Table 7.1 are real azimuths, but are angles bearing from north regardless of whether they were to the right or left of the meridian line. For this reason, and because our interest is to compare these values with those of the *decumani*, we have used the azimuths shown in Fig. 7.4b to compute the declinations, within [45°, 135°], which are equivalent to those shown in Fig. 7.5 and listed in Table 7.1.

For example, the value of maximum XII (Fig. 7.5a, Table 7.1), which is centred on 18° (α), approximates to that of the proportion 1:3. Due to the symmetry regarding cardinal points, this fraction comprises peaks II and VII in Fig. 7.4b, which derive from 3:1 and 1:3, respectively. In this way, the azimuths considered to relate to the declination were 72° (90° $-$ α) and 108° (90° + α), that resulted in $\delta \approx \pm 13.5°$ at latitude 40° and $\delta \approx \pm 13.25°$ at latitude 42°. In the map shown in Fig. 7.2 there is a concentration of cities that bear these azimuths in northern *Hispania*, while the rest are spread in the south, with just one in the Levantine coast.

Following the same criteria, maximum XIII (Fig. 7.5a), which is centred on 31.5° (Table 7.1), is associated with simple proportions such as 3:5, as well as 5:8, 7:11 and 7:12, and comprises the symmetrical maxima I and VII in Fig. 7.4b. The resultant declinations are $\delta \approx \pm 23.5°$ at latitude 40° and $\delta \approx \pm 22.75°$ at latitude 42°. Sites with this orientation are spread throughout the entire Iberian territory.

The last maximum analysed is XIV (Fig. 7.5), whose angle can be associated with *uariationes* 3:4 and 4:5, which are relatively simple, as well as sides of the Pythagorean triple 3:4:5. The resultant declination does not fall within a solar range but in a lunar one. The declinations are now $\delta \approx \pm 27.25°$ at latitude 40° and $\delta \approx \pm 26.5°$ at latitude 42°.

As can be observed from the histograms in Fig. 7.6, all of the formerly mentioned declination values correspond to the most significant maxima present in the studied sample. The declinations of peak XII approximate maxima B, E, H and L in Fig. 7.6, whose values coincide with that of the Sun on last days of April and October. Peak XIII gives solstitial or almost solstitial declinations and, as already mentioned, the resultant values for peak XIV offer declinations out of the solar range. To represent the histograms in Fig. 7.6 we have used an Epanechnikov kernel with a bandwidth of 1.5°, following the same criteria that we used in Figs. 7.4 and 7.5. Interestingly, the *uariationes* that fit the values of maxima X and XI, which involve higher proportions than those of XII, XXX and XIV (Table 7.1), give the less significant declinations groups of Fig. 7.6: C, D, J and K.

Finally, the same procedure was followed with the triangles in relation to the lowest Pythagorean triples: 3:4:5 and 5:12:13 if they were used as proposed in Fig. 7.1(2ab); that is, if the hypotenuse was drawn along the meridian line. For the triple 3:4:5 the resultant declinations were $\delta \approx \pm 27.25°$ at latitude 40° and $\delta \approx \pm 26.5°$ at latitude 42°. In the case of using 5:12:13, the results were

$\delta \approx \pm 17°$ at latitude 40° and $\delta \approx \pm 16.5°$ at 42°. As expected, values from 3:4:5 give the same declination as maximum XIV. The resulting declinations by using a triple 5:12:13 coincide with positions of the Sun on the first days of February and August and in mid-November and May.

Discussion

The fact that some azimuths in the sample give declinations for a flat horizon similar to the real ones in the Hispanic sample is not a surprise but something that one may expect. It is obvious as well that almost all the azimuths matched the *uaratio* angles, due to the high density of *uariationes* within an interval of 45° wide. However, the striking result is that the main declination groups of the sample (Fig. 7.6) coincide with significant dates when the Sun had azimuth values that resulted from the use of low *uariationes* (maxima XII, XIII and XIV). Furthermore, these declination groups are the same ones that contain the most significant astronomical relationships for the Roman context in *Hispania* (González-García et al. 2014).

In the first case, we observed that the fraction 1:3 (peak XII, Fig. 7.5) gives declinations at the studied latitudes that can be roughly associated with the position of the Sun on the day of the mythical foundation of Rome (21 April) or 1 May (as well as 1 November and 1 February for its negative value). In the case of 21 April, the connection with the Roman tradition is certainly obvious, but for the other dates, a reasonable explanation arises from pre-Roman traditions. In particular, these can be related to the so-called Celtic mid-season festivals (García Quintela and González-García 2017). This hypothesis makes sense for the group of cities in the Iberian northwest that bear azimuths from group XII (Fig. 7.5a), for being a region traditionally considered to have been inhabited by local societies with a Celtic substratum. In addition, similar results had been found in other regions with ancient Celtic traditions, such as *Galia* (García Quintela and González-García 2016; González-García and García Quintela 2014; González-García et al. 2016) or *Germania* (Espinosa Espinosa et al. 2016). We thus have found simple *uariationes* that relate to relevant astronomical orientations.

The second interesting result is that of peak XIII, which gives roughly solstitial declinations that are also present in the histograms in Fig. 7.6, and in the orientation patterns of Roman towns studied in other areas of the Empire (González-García and Magli 2015; González-García et al. 2018; Rodríguez-Antón et al. 2016b), that were explained by the symbolism of Augustus's propaganda, among other possibilities. In fact, 50% of towns that bear this azimuth have Augustan foundations or re-foundations: *Petavonium*, *Pisoraca*, *Valeria*, *Liberalitas Iulia* and the Imperial urban grids of *Corduba* and *Tarraco*. Once again, low proportions, such as the 3:5, give conspicuous astronomical patterns.

Finally, the use of the fraction 3:4 could have been applied to obtain orientations outside the solar range, like in the two maxima at $\delta \approx \pm 27°$ present in the declination histogram (F, G in Fig. 7.6). One hypothesis connects this result with the Julio-

Claudian dynasty, because the goddess Venus was regarded as the ancestral mother of the *Gens Iulia*, which Cesar and Augustus belonged to. There are three cities from this period with an azimuth of 3:4:5: *Juliobriga*, *Cesaraugusta* and *Barcino,* all in northern *Hispania* (Fig. 7.2). Similar values were also obtained in some Roman towns in North Africa (Rodríguez-Antón et al. 2017). In this particular case, some were explained by positions of the Moon and Venus as well, having also *Uthina* and *Thuburbo Majus* Augustan foundations (see González-García et al. 2018). However, it is just an hypothesis and a deeper analysis of these values is needed.

It must be stressed that not all the cities grouped in a declination pattern have the same azimuth, since the horizon is not always flat. There are also sites oriented according to some of the listed *uaratio* angles, for example 3:5, that have different declination values from those predicted for the selected latitudes and flat horizons. One example is that of the Roman theatre at *Carthago Nova* (present-day Cartagena), where the *frons scenae* and *cavea* line follows a solstitial azimuth that could have been drawn by the application of proportion 3:5 elsewhere or transported from the forum of the city (González-García et al. 2015), but the structure itself does not point to the real sunrise at the summer solstice due to the local topography.

In particular, just 57% of the cities in the declination group E have an azimuth that could be obtained by using a 1:3 relation and 66% of those with a solstitial declination have an azimuth corresponding to a fraction 3:5.

In the case of the relation 3:4, it is present in six cities with almost solstitial orientations or close to the extreme declination of Venus as a morning star ($\delta \approx \pm 24°$), and in eight towns where the *decumani* point towards extremes of Venus as an evening star ($\delta \approx \pm 27°$).

As a matter of fact we find sites with astronomical declinations and with azimuths derived from low *uariationes,* but there are also cities with the same declination but different azimuth values. In this last case, orientations might have been obtained by direct observation of astronomical phenomena. One example would be the orientation of Via Appia, where it seems that the Romans chose the setting of the star Castor as the direction where the via should face instead of the use of a simple fraction (1/1) (Magli et al. 2014). In this scenario, it is difficult to discern in cases where:

1. Direct astronomical observations give similar azimuths.
2. The orientations considered as astronomical resulted from the application of specific *uariationes*.
3. Standardised relations between *uariationes* and astronomical positions did exist so those fractions were used when direct observations of the astronomical phenomena were not possible.

Conclusions

In this analysis we have explored whether a standardised geometrical technique was used by ancient Roman topographers to set specific directions of land plots and, in particular, of city grids. By this, we can enumerate a couple of main outcomes:

1. The coincidence of the main minima of both distributions, *uaratio* angles and azimuths of the sample, suggests that there is a lack of orientations towards directions that are not associated with a *uaratio* proportion. Furthermore, the result of the Kolmogorov-Smirnov test does not allow us to refute the null hypothesis that both datasets come from the same parent population. That is, it is not possible to discard the idea that the azimuths were obtained by the use of the *uaratio* with a sufficient level of confidence.
2. Secondly, the azimuth groups of the sample that can be obtained by the application of low *uariationes* (XII, XIII and XIV) and the Pythagorean triple 3:4:5 give declination values that coincide with the main declination groups of the sample as well as with interesting astronomical positions within a Roman context. That is, there is an agreement between simple *uariationes* and relevant astronomical orientations present in the sample in *Hispania*. Conversely, maxima X and XI fit the angles of higher *uariationes* such as 1:9 and 2:9, and currently the resulting declinations cannot be explained by the relevant positions of any celestial bodies.

In the light of this, although sometimes orientations seem to hint an astronomical use of the *uaratio*, to better understand these results, an analysis of a wider sample at different latitudes, or restricted to delimited areas, is needed. Besides this, it would be also interesting to examine more local or chronological conditions, in order to determine whether astronomical motivations may have been favoured over more practical ones.

To sum up, it could be said that currently we cannot refute the use of a geometrical technique such as the *uaratio* to explain the orientation of Roman rural and urban spaces. In fact, this procedure could have been something extremely useful to obtain the characteristic regular organization of Roman landscapes according to the sky, since only the determination of the meridian line would have been needed to repeat the same orientation across the different territories. Furthermore, if a connection between the *uariationes* and the positions of some celestial bodies did really exist, geometry and astronomy might have represented a combination of practicality and symbolism when direct observations of the celestial phenomena in question were not possible.

Acknowledgments We express our gratitude to those who collaborated in the elaboration of this study: Esther Chávez-Álvarez, Luis Moranta, Elena H. Sánchez, Mario Gutiérrez and Aránzazu Medina. We also thank David Martín Belda for fruitful discussions. This work has been financed under the framework of the projects P/310793 'Arqueoastronomía' of the IAC, and AYA2015-66787 'Orientatio ad Sidera IV' and HAR2013-41635-P, of the Spanish MINECO.

References

Adam, J. P. (1982). Groma et chorobate. Exercices de topographie antique. *Mélanges de l'École Française de Rome, Antiquité, 94*, 1003–1029.

Bouma, J. (1993). *Marcus Iunius Nypsus. Fluminis varatio, Limitis repositio. Introduction, text, translation and commentary*. Frankfurt/Main: P. Lang (*Studien zur Klassischen Philologie*, 77).

Castillo Pascual, M. J. (1996). *Espacio de orden: El Modelo Gromático-Romano de Ordenación del Territorio*. Logroño: Universidad de La Rioja.
Chouquer, G. (2004). Une nouvelle interprétation du Corpus des Gromatici Veteres. *Agri Centuriati, 1*, 44–56.
Costa, V., & Orfila, M. (2014). Creación física de gnómones. In M. Orfila, E. Chávez-Álvarez, & E. H. Sánchez (Eds.), *La Orientación de las Estructuras Ortogonales de Nueva Planta en Época Romana* (pp. 129–155). Granada: Universidad de Granada-Universidad de La Laguna.
Espinosa Espinosa, D., González-García, A. C., & García Quintela, M. V. (2016). On the orientation of Roman towns in the Rhine area. *Mediterranean Archaeology and Archaeometry, 16*(4), 233–240.
García Quintela, M. V., & González-García, A. C. (2016). De Bibracte à Augutodunum: Observations archéoastronomiques. *Revue Archaeologique del'Est, 65*, 283–297.
García Quintela, M. V., & González-García, A. C. (2017). Archaeological footprints of the "Celtic calendar"? *Journal of Skyscape Archaeology, 3*(1), 49–78.
Gilman Romano, D. (2003). City planning, centuriation and land division in Roman Corinth: Colonia Laus Iulia Corinthiensis & Colonia Iulia Falvia Augusta Corinthiensis in Corinth, 20, Corinth, The Centenary: 1896–1996. *The American School of Classical Studies* (pp. 279–301).
González-García, A. C., & Costa Ferrer, L. (2011). The diachronic study of orientations: Mérida a case of study. In C. L. N. Ruggles (Ed.), *Archaeoastronomy and ethnoastronomy: Building bridges between cultures* (pp. 374–381). Cambridge: Cambridge University Press.
González-García, A. C., & García Quintela, M. V. (2014). The 1st of August at Lugdunum: Astronomy and imperial cult in Galia. *Mediterranean Archaeology and Archaeometry, 14*(3), 83–91.
González-García, A. C., & Magli, G. (2015). Roman city planning and spatial organization. In C. L. N. Ruggles (Ed.), *Handbook of archaeoastronomy and ethnoastronomy* (pp. 1643–1650). Heidelberg: Springer.
González-García, A. C., & Šprajc, I. (2016). Astronomical significance of architectural orientations in the Maya lowlands, a statistical approach. *Journal of Archaeological Sciences: Reports, 9*, 191–202.
González-García, A. C., Rodríguez-Antón, A., & Belmonte, J. A. (2014). The orientation of Roman towns in Hispania: Preliminary results. *Mediterranean Archaeology and Archaeometry, 14*(3), 107–119.
González-García, A. C., Noguera Celdrán, J. M., Belmonte Avilés, J. A., Rodríguez-Antón, A., Ruiz Valderas, E., Madrid Balanza, M. J., Zamora, E., & Bonnet Casciaro, J. (2015). Orientatio ad sidera: Astronomía y Paisaje urbano en Qart Hadašt/Carthago Nova. *Zephyrus, 75*, 141–162.
González-García, A. C., García Quintela, M. V., & Belmonte, J. A. (2016). Landscape construction and time reckoning in Iron Age Celtic Iberia. *Documenta Praehistoria, 43*, 479–497.
González-García, A. C., Rodríguez-Antón, A., Espinosa Espinosa, D., García Quintela, M. V., & Belmonte, J. A. (2018). Establishing a new order: The orientation of Roman towns at the age of Augustus. In G. Magli, A. C. González-García, J. Belmonte Aviles, & E. Antonello (Eds.), *Archaeoastronomy in the Roman world*. Heidelberg: Springer.
Gros, P. (1976). Nombres irrationnels et nombres parfaits chez Vitruve. *Mémoires de l'Ecole Française de Roma, Antiquité, 88*(2), 669–704.
Le Gall, J. (1975). Les Romains et l'orientation solaire. *Mélanges de l'École Française de Rome, Antiquité, 87*, 287–320.
Lewis, M. J. T. (2001). *Surveying instruments in ancient Greece and Rome*. Cambridge: Cambridge University Press.
Magli, G., Realini, E., Reguzzoni, M., & Sampietro, D. (2014). Uncovering a masterpiece of Roman engineering: the project of Via Appia between Colle Pardo and Terracina. *Journal of Cultural Heritage, 15*, 665–669.
Neugebauer, O. (1969). *The exact sciences in antiquity*. New York: Dover.
Orfila, M. (2012). Un posible sistema para orientar estructuras de trazado ortogonal en época clásica. El caso de Pollentia (Mallorca, España). *Geographia Antiqua, 20–21*, 123–136.

Orfila, M., & Moranta, L. (2001). Estudio del trazado regulador del Foro de Pollentia (Alcúdia, Mallorca). *Archivo Español de Arqueología, 74*, 209–232.

Orfila, M. (2014). Los triángulos en el mundo antiguo. In M. Orfila, E. Chávez-Álvarez, & E. H. Sánchez (Eds.), *La orientación de las estructuras ortogonales de nueva planta en época romana. De la uaratio y sus variaciones* (pp. 67–69). Universidad de Granada – Universidad de La Laguna: Granada.

Orfila, M., Chávez-Álvarez, E., & Sánchez, E. H. (2014). *La Orientación de las Estructuras Ortogonales de Nueva Planta en Época Romana. De la Varatio y sus Variaciones*. Granada: Universidad de Granada-Universidad de La Laguna.

Orfila Pons, M., Chávez-Álvarez, M. E., & Sánchez, E. H. (2017a). Fundaciones en época Romana. De lo intangible a lo tangible. ¿Cuándo, por qué, dónde, cómo, simbología? *Sagvntvm-PLAV, 19*, 265–276.

Orfila Pons, M., Chávez-Álvarez, M. E., & Sánchez, E. H. (2017b). Urbanizar en época Romana: Ritualidad y practicidad. Propuesta de un procedimiento homologado de ejecución. *SPAL, 26*, 113–134.

Peterson, J. W. M. (1992). Trigonometry in Roman cadastre. In J. Y. Guillaumin (Ed.), *Mathématiques dans l'Antiquité* (pp. 185–203). St.-Étiene: Université de St-Étienne.

Peterson, J. W. M. (2007). Random orientation of Roman camps. *Oxford Journal of Archaeology, 26*(1), 103–108.

Richardson, A. (2005). The orientation of Roman camps and forts. *Oxford Journal of Archaeology, 24*(4), 415–426.

Rodríguez-Antón, A. (2017). *Cosmovisión y Urbanismo en la Roma Antigua: Orientación de Ciudades y Campamentos Romanos*. Ph.D. thesis. Universidad de La Laguna-Instituto de Astrofísica de Canarias, Spain.

Rodríguez-Antón, A., González-García, A. C., & Belmonte, J. A. (2016a). Orientation of Roman camps and forts in Britannia: Reconsidering Alan Richardson's work. In F. Silva, K. Malville, T. Lomsdales, & F. Ventura (Eds.), *The materiality of the Sky* (pp. 135–144). Ceredigion: Sophia Centre Press.

Rodríguez-Antón, A., Belmonte, J. A., & González-García, A. C. (2016b). Romans in the Near East: The orientation of Roman settlements in present-day Jordan. *Mediterranean Archaeology and Archaeometry, 16*(4), 153–160.

Rodríguez-Antón, A., González-García, A. C., & Belmonte, J. A. (2017). Estimating the reliability of digital data acquisition in cultural astronomy. Its use in the case of Roman North Africa. *Culture and Cosmos, 3*(2).

Roskams, S. (2001). *Excavation: Cambridge manuals in archaeology*. Cambridge: Cambridge University Press.

Roth-Congés, A. (1996). Modalités pratiques d'implantation des cadastres Romains: Quelques aspects (Quintarios claudere. Perpendere. Cultellare. Varare: la construction des cadastres sur une diagonale et ses traces dans le Corpus Agrimensorum). *Mélanges de l'Ecole Française de Rome, Antiquité, 108*(1), 299–422.

Rykwert, J. (1988). *The idea of a town. The anthropology of urban form in Rome, Italy and the ancient world*. Princeton: Princeton University Press.

Part IV
Roman Provinces

Chapter 8
Arabia Adquisita: The Romanization of the Nabataean Cultic Calendar and the Tannur 'Zodiac' Paradigm

Juan Belmonte Aviles, A. César González-García, and Andrea Rodriguez-Antón

Abstract One of the most fascinating and enigmatic pieces of evidence of Nabataean ingenuity is the so-called Zodiac of Khirbet et-Tannur (Jordan), found in a temple built at the mountain summit close to Djebel Tannur in the first half of the second century CE, possibly when the ancient Nabataean Kingdom was already under Roman rule. However, Nabataean traditions and cults persisted during the Roman period and even survived well into Byzantine times. But one important change was the imposition of a Julian-like calendar, of Egyptian inspiration, instead of the original lunisolar calendar of the Nabataeans—earlier inherited and adapted from the Babylonian one—whose month names were however preserved under a solar perspective and a new time framework entitled *Era Provincia Arabia*. An analysis of the dates reported in the foundation inscriptions of the first century CE rock-carved tombs at the southern Nabataean city of Hegra, and other dated inscriptions of the Nabataean Kingdom period, has given some clues for us to look at the Tannur Zodiac with a different perspective. This new way of thinking has allowed a completely different approach to this masterpiece of art which is substantially different to most previous interpretations (see e.g. McKenzie et al., ARAM Periodical 24:379–420, 2012). According to our hypothesis, we consider that it should be formally named the 'almanac' or 'parapegma' of Khirbet et-Tannur hereafter.

J. B. Aviles (✉) · A. Rodriguez-Antón
Instituto de Astrofísica de Canarias and Universidad de La Laguna, La Laguna, Tenerife, Spain
e-mail: jba@iac.es; ara_ext@iac.es

A. C. González-García
Instituto de Ciencias del Patrimonio, Incipit-CSIC, Santiago de Compostela, Spain
e-mail: a.cesar.gonzalez-garcia@incipit.csic.es

G. Magli et al. (eds.), *Archaeoastronomy in the Roman World*, Historical & Cultural Astronomy, https://doi.org/10.1007/978-3-319-97007-3_8

Introduction: A Nabataean Cultic Calendar?

In March 106 CE the Nabataean Kingdom was annexed by Emperor Trajan to the Roman Empire creating the new province of Arabia (Petraea). Ancient Nabataea, and its ancient capital, the rose city of Petra, has been one of our main research objectives since our first field campaign in the region in 1996 (Belmonte 1999). In December 2015 a new visit to the area was arranged that coincided with the Winter Solstice. Several illumination effects were observed and broadcasted at the principal monuments of Petra and new important hierophanies, predicted in previous campaigns (Belmonte et al. 2013), were verified and contrasted in the light of the literary and epigraphic sources and astral symbolism (Belmonte and González-García 2017). This allowed us to conduct a new analysis of every single clue of astronomical relationships in ancient Nabataea, including all evidence in their time-keeping system.

The idea of a Nabataean calendar centered on the cult of their deities and their ancestors in certain time-marks of special astronomical significance along the year has recently been developed. This has included a combined analysis of classical historiography, ethnographic sources, epigraphy and the archaeological record, interpreted in the light of cultural astronomy (Belmonte and González-García 2017). This was framed within the Babylonian lunisolar calendar of the Seleucid Empire, but not fully restricted by it since the Nabataean calendar seems to have started in Nisan, although most of the calendars of Asia Minor and the Near East in the Hellenistic and Roman Periods began the year in autumn (Stern 2012). According to this proposal, the principal astronomical milestones of this calendar would have been:

1. New Year's Eve on 1 Nisan. This could happen either before or after the Spring Equinox (see below).
2. The Full Moon after Spring Equinox (14 Nisan), in a clear parallelism to Jewish Passover.
3. The First Crescent after Spring Equinox (1 Nisan—New Year's Eve—or 1 Iyyar otherwise)
4. With some doubt, since evidence is not strong: the First Crescent of the lunar month including Summer Solstice (1 Tammuz).
5. The First Crescent after Summer Solstice (1 (B)ab)
6. The heliacal rising of Sirius (Isis) in c. 20/7 Julian. Hence in (B)ab. This could be related to the cult of Isis-Al Uzza as clear evidence of Hellenistic Egypt influence.
7. The First Crescent after Autumn Equinox (1 Tishri), not reflected, however, in the epigraphy.
8. The pilgrimage festival of Aggathalbaeith in the first Lunar Month after the Autumn Equinox (most likely Tishri).
9. The Winter Solstice and the Birth of Dushara, as celebrated during the Epiphany in late Roman times.

10. The First Crescent or Full Moon of Thebet, the lunar month after—or including—the Winter Solstice.

These dates would have been the times for the main festivals and celebrations which on at least three occasions—the two equinoxes and the Winter Solstice—would have been performed in the form of ritual pilgrimages to attractive sacred spots in Nabataea and indeed in the capital city of Petra. These spots included Ad-Deir (and the Lions' Triclinium), the high-places at Jabal Madbah and Jabal Khubza and the summit of Jabal Hārūn, all of them in Petra and its vicinity, or the imposing temple of Khirbet et-Tannur, further north, which will be the nucleus of this essay.

As argued, the Nabataean Kingdom had a lunisolar calendar of Babylonian type inherited during the Seleucid period (Alzoubi 2016), but in CE 106 the Kingdom was annexed by Emperor Trajan. As implied by Babatha's Archive, a new era was soon initiated 'according to Provincia Arabia' with New Year's Eve at 1 Nisan/Xandikos, corresponding to 22 March in the Julian Calendar and with a fully solar structure (length of 365 days with a bissextile after each 4 years, Samuel 1972). Nabataeans needed to adapt these lunar dates to these new circumstances, and they did it very well as the analysis of Khirbet et-Tannur 'Zodiac' will show.

Petra, the ancient Nabataean capital, has been one of our main research objectives since the first field campaign at the site in 1996 (Belmonte 1999). In December 2011, a large field campaign was mounted at sites of the Nabataean Kingdom belonging to nowadays Jordan, including Khirbet et-Tannur and Khirbet edh-Dharih, which offered further clues to the relevance of cultural astronomy in sacred building design and orientation (Belmonte et al. 2013). Last, but not least, in December 2015, a new visit to Petra was made to coincide with the Winter Solstice. Historical, ethnographic, epigraphic and archaeological records have been deeply analysed and compared in order to gain an insight into the Nabataean land- and skyscape relationship. From this multidisciplinary analysis two main points were evident: (1) the importance of both equinoxes and Winter Solstice as time-markers within the lunisolar calendar, and (2) the relevance of some processions and pilgrimages. One of these examples is illustrated in Fig. 8.1.

Illumination effects have been observed and recorded at the principal monuments of Petra. These strongly suggest the relevance of these dates at the time of the Nabataeans. The Winter Solstice was an important event in the Nabataean cultic calendar when a festival of the main deities of the city, the god Dushara and his partner the goddess Al-Uzza, was commemorated. This probably took the form of a pilgrimage, and related cultic activities, such as ascending from the temples at the centre of the city (presumably from Qasr el Bint and the Temple of the Winged Lions) to the Monastery (Ad-Deir) through an elaborated stone-carved processional way (see Fig. 8.1 and below). The relevance of spring and the Spring Equinox within the cultic calendar is also emphasized in relation to other sacred sites in Petra, such as

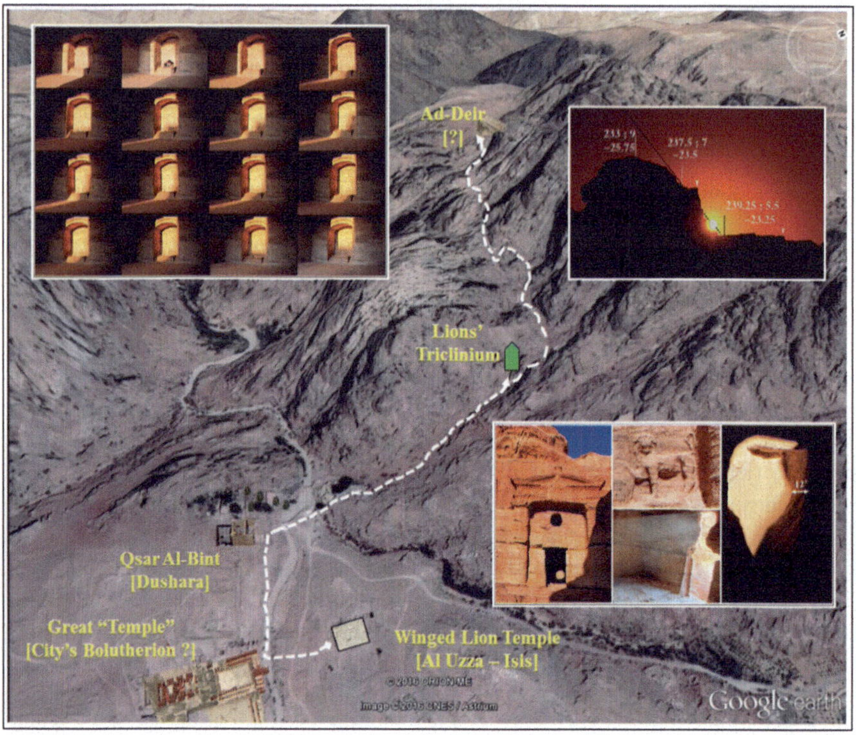

Fig. 8.1 The proposed pilgrimage route—mostly carved out of the rock—from the city centre (temples of Al-Uzza and Dushara) up to the mountain plateau where the Monastery (Ad-Deir) was sculpted on the sandstone cliffs. The route passed by the Lions' Triclinium where a nice light and shadow effect happened at Winter Solstice sunrise (below). Winter Solstice sunset light and shadow effects happened both in the interior of Ad-Deir and at the horizon as seen from it (upper left and right respectively). This could be the most important Nabataean pilgrimage route in Petra relating to Dushara's birth festival from his mother *cum*-consort Al-Uzza (diagram by the authors upon an image courtesy of Google Earth)

the Zibb Atuff obelisks, and additional Nabataean sites (Belmonte and González-García 2017). It is worth explaining in detail one such pilgrimage route, but before we do this a short reminder about Nabataean religion and related ethnohistorical sources is certainly warranted.[1]

[1]This information has already been published in Belmonte and González-García (2017) but for the sake of clarity and completeness of the present essay we consider that it is appropriate to repeat it here.

Inspiring Astral Deities in the Nabataean Realm and Their Festivals

The Nabataeans had a religion that was a strange mixture of elements from pre-Islamic Arabs and Hellenistic, Egyptian and Middle Eastern influences. Divinities were often represented by stone blocks or betyls (*neshebet* in Nabataean) although in the late period and notably under Roman rule, human or quasi-human forms were developed (see Healey 2001, for a thorough discussion on this topic). The main male divinity was the god Dushara, or Dushares, most likely an astral god. His name means 'He of Shara', Shara being the mountain range bordering Petra to the east where the neighbourhood of Gaia, or Al-Ji (today Wadi Musa) was located. The area of Al Madras, to the south of Wadi Musa was especially sacred to him. On certain occasions, he seems to be a form of the god Al-Kutba ('the one who writes'), although this is much debated. Dushara was identified by classical writers either as Zeus, Ares or Dionysos. Triclinium no. 17 in Bab es-Siq at Petra has an inscription dated 96/95 BCE devoted to Dushara: this is the oldest dated Nabatean inscription and probably the oldest dated betyl of the god (Wenning 2001). According to Suidas [Theus Ares]

> ... the god Dushara is worshipped by them for him they honour above all others. The image is a black stone square and unshapen, four feet high by two feet broad—one foot in thickness. It is set on a base of wrought gold. (Hoyland 2001: 183).

Dushara's presence in Nabataea both during the Kingdom and under Roman rule (*Provincia Arabia*) is overwhelming.

On the contrary, there has been much discussion regarding the head female divinity of the Nabataean pantheon (Healey 2001). In Bosra, the northern Nabataean capital during the reign of Rabel II (CE 71–106) and the capital of the Roman province, the main goddess was Allat (or Al Lat), meaning simply 'The Goddess'. With a hypothetical solar character (*As Sams*, the Sun, was a female divinity in pre-Islamic Arabia), she has been identified with Athena and Atargatis (the Syrian Goddess). Her name has also been found in Palmyra (Syria), in Iriam (Wadi Rum, Jordan), and in Hatra (Iraq) and she is indeed mentioned in the Quran. She is identified with the Alilat of Herodotus (History III.8).

Interestingly, in Petra, this name is never found. Instead, inscriptions refer to the goddess Al-Uzza. Her name means 'The Most Powerful' and she was the personification of the planet Venus—often specifically assigned to the Evening Star (Hawting 1999), identified with Aphrodite and the Cananaean Astarte and also with the Egyptian goddess Isis (Merklein and Gerger 2001). In inscriptions in Petra and Iriam (present day Wadi Rum), she is mentioned together with Al-Kutba, the 'Master of the House', and of course Dushara, of whom she could have been his partner or even his mother-consort. It has been argued that Al-Uzza was possibly an alternative manifestation of Allat—e.g. the lion was the totem animal for both deities. Interestingly, she seldom had a double nature, being known as Al-Uzzatan, so that, according to many scholars, Allat and Uzza could go in tandem, being two faces of a same coin

(Hawting 1999). However, the Holy Quran reports in Sura 53: "Have thou seen Al Lat, Al-Uzza and the other, Manat, the third one. These are only names that thou and thine ancestors had given to them. Allah has not put any power in them." Apparently, according to these verses, these three goddesses were different personalities, who have been identified as the daughters of Allah. However, this does not necessarily apply for Nabataean and even Roman times several centuries earlier. The controversy still continues (Gawlikowski 1990).

What seems to be a completely different female personality is Manatu, a deity (or deities, since the word is a plural) who joins Dushara in several funereal inscriptions at Hegra (Median Saleh, Saudi Arabia; Healey 1993). She was the goddess of fortune and the numen of that southern city and is usually believed to be Dushara's daughter. Manatu as Fate was conceived as a trio of goddesses in Antiquity (Hawting 1999). It is possible that the Moon was one of her manifestations and she could have been represented as three adjacent betyls in several niches across Petra and Hegra. Finally, stars were also very important in pre-Islamic Arabia and bright stars and asterisms such as Sirius (Sira'), Canopus (Suhail) and the Pleiades (An-Nijm) were broadly used for guiding the caravans, establishing the dates of pilgrimages, regulating calendars or making weather forecasts (Forcada Nogués 1993). This could have been so in ancient times. In this line of argument, it is worth mentioning that Nabataean Queens were often assimilated with Isis-Al-Uzza and that Sirius, in her name of Sopdet, was the main celestial aspects of Isis in ancient Egypt (Belmonte 2012).

The Nabataean deities are often mentioned in the writings of ancient sages of Antiquity. Strabo (Geo 16, 4, 26) informs us that the Nabataeans "... worship the sun, building an altar on the rooftop of their houses, pouring libations on it every day and burning frankincense." This sort of domestic cultic practice presumably had its official counterpart in the numerous high-places still standing in Petra and elsewhere. Some of them are of a size fit for a family group or clan; others have a much larger monumental character and often are located at mountain and hill tops, and possibly relate to important festivals. Among these, the outstanding temple at Khirbet et-Tannur can certainly be included.

Historical sources for cult practices in the Levant in the Hellenistic and Roman periods are scarce but most relevant. Lucianus (Dea Siria 49) reports that in the city of Hierapolis-Mambij, in northern Syria, "... the greatest festival they celebrate is that held in the opening of the spring." But it is in slightly later sources where information is paradigmatic. For example, Procopius (2, 16) informs us that the time for festivals "... was the season of the Vernal Equinox and at this season the Saracens always dedicated about 2 months to their god." The 'Saracens' was the name given by late Roman sources to the Arab-speaking tribes who inhabited the ancient lands of the Nabataeans and still used, among others, the Nabataean Aramaic alphabet for their inscriptions. Probably, they were somehow related to the Nabataeans of the Kingdom period, if not their direct descendants. This information can be further extended with the text of Protheus (3) who argued that the grammarian Nonnius reported that most of the Saracens gather at a certain sacred place—not explicitly mentioned—twice each year. The text continues:

> ... the first of these assemblies extends over a whole month and takes place about the middle of the spring, when the sun passes through the sign of Aries, while the other lasts two months; this they celebrate after the Summer Solstice.

The arrival of the New Moon marked the time for these festivals (Hoyland 2001). To judge from the evidence of pre-Islamic northern Arabia, annual spring festivals are likely to have been held with aspects of pilgrimage attached (Healey 2001).

Another clue to the identity of the Saracens and their close ties with the Nabataeans is that of the Christian apologist Jerome (Vita Hilarionis, 42–43) who reported that they

> ... arrived at Elusa on the very day that the solemn festival [which] had brought all the people of the town to the temple of Venus; for the Saracens worship this goddess as the Morning Star and their race is dedicated to her cult.

Early scholars already argued that the name of the city of Elusa was actually a Greek spelling for Al-Uzza and that the name of the city actually honoured the greatest Nabataean goddess, reinforcing the idea stated above. Indeed, festivals dedicated to the greatest female divinity were amongst the most important in the Levant.

In this line of argument, the texts of the Christian apologist Epiphanius of Salamis (fourth century CE, who was born at Eleutheropolis, modern Bayt Jibrin, a locality in ancient Edom and was hence a native of the region) are very important and among the most elucidating. This is so even considering the time passed from the Roman annexation of the Nabatean Kingdom and that his arguments were written to support his assertions and to show that the cult of the virgin had its pagan equivalences (Mordtmann 1875). Epiphanius reported that

> ... in the idolatrous temple at Petra ... they praise the virgin with hymns in the Arabic language and call her khaabu ... in Arabic; and the child who is born of her they call Dusares. And this is also done that night in the city of Elusa, as it is there in Petra ... on the very night of the Epiphany. (Contra Haeretici, Panarion, 51, 22).

Hence, according to this text, written ca. CE 374–376, the celebration of the birth of Dushara from the womb of the Virgin Mother (almost certainly his mother-cum consort Al-Uzza, or Allat if they are to be identified) took place on 6 January in Petra at dates close to the Winter Solstice, supporting Dushara's solar character. In the same line of argument, it has been suggested that

> ... all these evidences may well point towards a solar character of the Nabataean chief god and incline us to consider the 'very great festival on the very night of the Epiphany', mentioned in Epiphanius, as a survival of an earlier celebration of the Winter Solstice among the Nabataeans. (Janif 2006–2007: 345).

Lastly, Epiphanius also informs us that the fourth century Arab tribes of southern Palestine and Transjordan performed a pilgrimage during the month of Aggathalbaeith to a major sanctuary—'al baeith', The House—in the region (Panarion 51, 24). This could have been Petra, since Dushara is often called *mr' byt'*, Lord of the House in Nabataean inscriptions, or any other relevant sanctuary, perhaps Khirbet et-Tannur. The time of the pilgrimage corresponds to the month of

Tishri, the first lunar month after the Autumn Equinox in the Nabataean lunisolar calendar. The central day of the festival was on the 22nd day of Aggaathalbaeith, corresponding approximately to 8 November in the Julian calendar. This cult strongly resembles the later Muslim tradition of pilgrimage (*hajj*) in the month of Dhu al Hijja to the 'House of God' in Mecca, the Ka'aba (Janif 2006–2007).

Land- and Skyscape in Petra: The 'Winter Solstice' Pilgrimage

As clearly shown, Nabataeans certainly celebrated religious festivals and performed pilgrimages to sacred shrines at the time of these celebrations. However, direct evidence on the terrain has so far been very scanty if it were not for the imposing ascending routes to several sacred sites at various spots within the city of Petra (El Khoury 2006–2007; see e.g. Fig. 8.1) and elsewhere, certainly including Khirbet et-Tannur as we shall shortly demonstrate. The question is how, why and when these pilgrimage routes could have been used. It is perhaps useful to return to our earlier results (Belmonte and González-García 2017) and analyse in certain detail how, why and when one such route was possibly in action: the pilgrimage to Ad-Deir in Petra.[2]

Petra city centre is spread over a vast area of *circa* 1 km wide between the slopes of Jabal al Khubza and the cliffs of Umm al Biyara. Here, three large free-standing structures were erected by the Nabataeans in the first century CE. Among them are the Qsar al Bint, dated to ca. CE 40, and the Temple of the Winged Lions, dated to CE 27/28, presumably the temples of Dushara and Al-Uzza, respectively (Alpass 2010; Browning 1989). As shown in Fig. 8.1, one of the most important pilgrimage ascending routes of Petra was the one departing from the area of these temples, at the city centre, and marching up the mountains to a high plateau open to the western horizon where the most imposing monument of the city was built: Ad-Deir. Here a possible astronomical orientation related to the time of the Winter Solstice was discovered two decades ago (Belmonte 1999).

As the processional way goes up to Ad-Deir there are several attention foci. One of them is the Lions' Triclinium which is located in a narrow chasm beside the path (see Fig. 8.1). Nicely carved into the sandstone, it has a reworked niche for a cult statue in its back wall. Data obtained in 2011 (Belmonte et al. 2013) clearly predicted that an interesting light and shadow effect could be produced at the niche at the moment of the Winter Solstice thanks to the light entering across a—now worn—oculus in the façade. During the Winter Solstice of December 2015 the effect was observed in all its splendour at that precise moment. This hierophany is only produced at sunrise in

[2]Although fully described in Belmonte and González-García (2017), we believe that it is worth repeating this description so that the reader can get a first-hand impression of how a Nabataean pilgrimage route would have worked in the Kingdom period, and certainly later during the Roman dominion.

dates close to the Winter Solstice since the triclinium would be in darkness for the rest of sunrises throughout the year, as was experienced by McKenzie's team in late January when the effect was hardly visible (Mckenzie 2013). Still more, considering the variation in the obliquity of the ecliptic, the effect must have been still more spectacular in Nabataean times (see Fig. 8.1, extreme bottom right image). So, either on the way up during the early morning of the day of the pilgrimage, or on the way down at dawn, after a night of celebration at the Ad-Deir plateau, the procession could stop at the Lions' Triclinium to glimpse a nice spectacle, which certainly related to the rituals associated with the Winter Solstice and the birth of Dushara.

However, the final objective of the pilgrimage was obviously Ad-Deir. Is this the temple of one of the Nabataean divinities or the unfinished burial place or cenotaph of one of their last kings? Its use as a church in the Byzantine period and its internal structure strongly support Browning's (1989: 190–5) assessment: Ad-Deir was "... a prominent festival venue, with an elaborated staged ascent to it and a vast court in front of it." Its astronomical orientation and the light and shadow effects produced at the moment of the Winter Solstice both at the môtab for the cult betyls and at the nearby western horizon seem to ratify this line of argument (see Fig. 8.1). The Nabataeans worshipped the môtab (*mwtb'*), the podium on which the stele/betyl was erected, the equivalent of the seat or throne of the deity. Actually, slight traces of a rock-carved cult betyl, with Dushara's block proportions—possibly destroyed when the site was converted into a church—could still be discerned at the centre of the môtab. The light and shadow effect of the double sunset phenomenon produced at Ad-Deir reaches its maximum precision and beauty at this position precisely. Interestingly, a very important fact is that the vaulted niche with the môtab was carefully positioned slightly off-centre relative to the doorway (McKenzie 2013). This asymmetry allowed the effect to occur, which was hence included in the architectural design.

The hierophany is spectacular and would have been observable for nearly a week before and after the Winter Solstice. Winter Solstice sunset, as observed from the môtab itself, is produced in a peculiar way on a modified rock with the aspect of the head of a lion—the sacred animal of Al-Uzza. At the present time the Sun sets at least twice, first in the axis of the monument and then re-appears in the northernmost corner of the rock before its final disappearing. It should be noted that the anthropic modification of this rock in the area where this phenomenon occurs, as seen from the môtab of Ad-Deir, allows for such observation and it would also have been part of the design of the complex (i.e. a modified horizon). The phenomenon would have been still more impressive two thousand years ago when the northern limb of the disk of the Sun had a declination close to $-23\frac{1}{2}°$ (Belmonte et al. 2013).

Undoubtedly, this ensemble of solar hierophanies confirms the idea that the Monastery was one of the most important sacred enclosures of the Nabataean realm and that Petra was the goal of one of the most important pilgrimage routes, and this happened not only during the Nabatean Kingdom but also under Roman rule. Ad-Deir possibly was the ideal place to celebrate 'on the very night of the Epiphany'—in dates close to the Winter Solstice—the birth of Dushara from his own mother-cum-consort Al-Uzza, the goddess of fertility, reflected in Epiphanius's writings. Echoes of this ancestral ritual might still be alive in the area of Petra as shown by the surviving tradition of the *Amm al-Gaith* procession (see e.g. Al-Salameen and Falahat 2009; Belmonte and González-García 2017 and references therein). It is now the time to turn to the main objective of this essay, the temple of Khirbet et-Tannur and the noteworthy 'Zodiac' discovered at this site.

The Twin Temples of Tannur (Hurawa) and Dharih (Aïn Al-La'aban)

The deep gorge of the Wadi Al-Hasa marks the limits between the ancient lands of Edom and Moab. Precisely at the place where the King's Road (later followed by the *Via Nova Traiana* after the Roman annexation of Nabataea) passes close to an impressive topographic marker, the blackish stone behemoth of the Djebel Tannur—the Mountain of the Oven, the rest of the nucleus of a long-extinct volcano (see Fig. 8.2). Close to it, across the riverbed to the south, there is a high hill that also constitutes a landmark in the territory (Fig. 8.2, panel a). On the very top of that hill a huge, noteworthy and conspicuous shrine was built: the so-called temple of Khirbet et-Tannur. The sanctuary was fully excavated in the 1930s by Nelson Glueck and his team and the extraordinary findings, including sculptures and high-reliefs, were brought down the mountain and distributed between the Museums of Amman and Cincinnati (the best pieces) where they can still be admired today (unfortunately completely out of context). Glueck never published the excavations in full and eight decades have elapsed until the extraordinary work of Judith McKenzie and her team has brought back the temple, at least on paper, to all its lost glory (see McKenzie 2013).

McKenzie's seminal work offers a complete description of the findings and a full-scale reconstruction of how the temple and its component should have looked like. It is perhaps worth summarizing, following a diachronic sequence, some of her team's major discoveries which are relevant for our purposes:

1. The earliest dated objects on site are two Seleucid coins of late third and early second centuries BCE which suggest the earlier construction of certain structures in the second century BCE on this site.
2. The discovery of an aniconic stele/betyl dated in the first century BCE dedicated in Nabataean to the Edomite god Qosh at Hurawa (the proper name of the site?). The god Qosh was the main Edomite deity and his presence here possibly speaks of a much earlier sacred character of the mountain.
3. In the year 8/7 BCE, there is a dedication of a building (Period 1) on the site by the Head of Aïn al-La'aban, presumably the nearby site of Khirbet edh-Dharih (see Fig. 8.2).
4. The finding of a typically Nabataean bowl and pottery shows that the use of the site noticeably increased in the first century CE.
5. In the first half of the second century CE, the whole complex was rebuilt on a monumental scale (the so-called Period 2). However, it is not clear whether

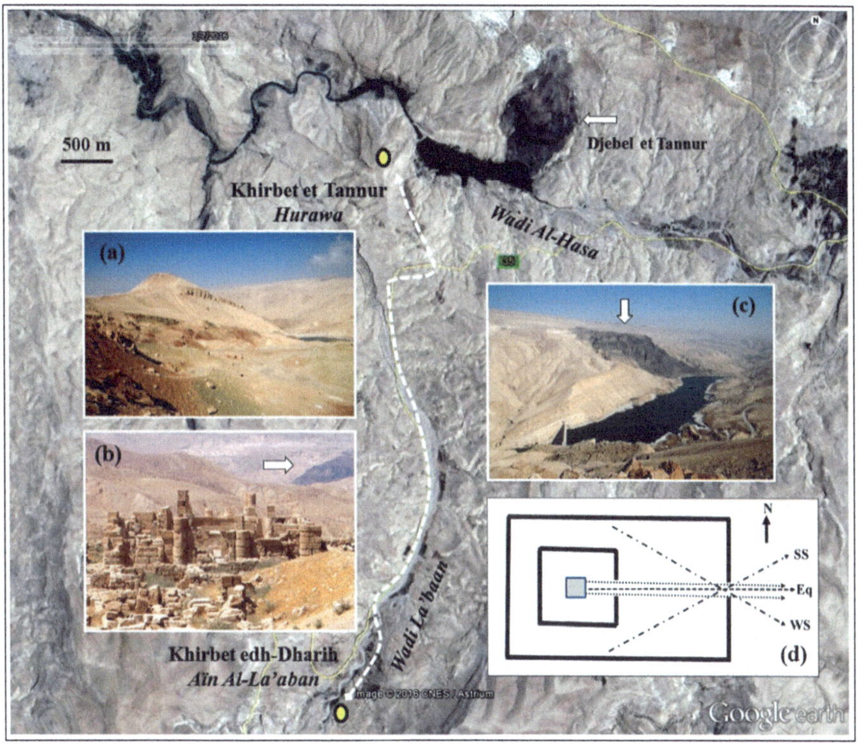

Fig. 8.2 The region of Wadi Al-Hasa north of Petra, showing the proposed pilgrimage route that would have been followed by pilgrims to the temple of Khirbet Tannur, possibly ancient Hurawa, located at the top of a hill south of wadi Al-Hasa (**a**). They eventually departed from the main temple at Khirbet edh-Dharih (Aïn Al-La'aban, **b**). The Djebel Tannur (**c**) is the dominant topographic feature of the area but neither of the two temples was orientated in its direction. Hurawa was, however, orientated towards New Year's Eve sunrise of the *Arabia* province calendar in Roman times (**d**). See the text for further details (diagram by the authors)

construction began before or after the Roman annexation of Arabia in CE 106. Interestingly, the enclosure wall at Khirbet Edh-Dharih has a *terminus post quem* in CE 156/156 provided by a coin of Emperor Antoninus Pius found in its foundations.

6. The 'Zodiac' was discovered in the sancta sanctorum of the temple. It has a peculiar distribution of zodiacal signs (see Fig. 8.3) and belonged to the late phases of this period of construction when the impressive decorative programme of the site was carried out.
7. The plan is unique for a Nabataean temple because of the tall altar platform, with a niche in the front for the statues and steps to the altar on top. Actually, it seems to reproduce in stone a typical rock-carved Nabataean high-place.
8. The temple faces East, but "It is unusual for temples to face East so precisely ..." according to McKenzie (2013: 217). However, our data here and elsewhere

Fig. 8.3 The 'Zodiac' of Khirbet Tannur, encircling a Nabatean goddess, presumably the main female deity of the couple worshiped at the hill-top temple. The twelve identified zodiacal signs follow a non-standard ordering dividing the circle into two halves. Aries and Capricorn are personified images perhaps related to certain Nabataean deities. The position of the five dots has been highlighted. See the text for further details (diagram by the authors)

in Nabataea suggest that this was not so uncommon a practice (Belmonte et al. 2013; Rodríguez-Antón et al. 2016).

9. Evidence of feasting and night-time activity have been found. This fact is most relevant for our interests.
10. The platform altar was enlarged in a second phase later on (Period 3). Six busts in two columns of personifications of the zodiac were added to the altar (Virgo: Grain Goddess, and Pisces: Fish Goddess the only ones nicely preserved). Their relative position is identical to the 'Zodiac', down below left and right, respectively. This implies a sort of continuity from Periods 2 to 3 and that the order of signs is deliberate and not the result of a Nabataean misunderstanding of classical iconography. This is an important point.
11. Finally, the temple was destroyed by the CE 363 earthquake, which was accompanied by a fire that caused so much destruction that worship at the site largely ceased. At that stage, the Zodiac stela was broken and the main fragments were buried under piles of rubble, which contributed to their marvelous

state of preservation. They would be separately discovered later on within an interval of 20 years, contributing to the mystery of its interpretation.

The temple was devoted to a divine couple, a seated god and goddess, framed by bulls and lions respectively, with the Zodiac-Nike slab—described below—presumably located between and above them according to the most probable reconstructions (McKenzie 2013). We have seen that in the first century BCE the god of the sanctuary was Qosh. He was a supreme storm-god, producing rain to ensure a plentiful harvest, good pasture and herds. The surviving cult image of the Period 2 temple (second century CE) is that of a male deity with attributes of Hadad, Zeus and Serapis and hence of a supreme god. Since the supreme Nabataean god was Dushara, the god of the cult statue was possibly a form of this god with the attributes of a storm-god: namely Qosh-Dushara. However, the identification of the female deity is not so clear because few fragments have been preserved. She would apparently be a form of Atargatis but, on the one hand, Allat had a leonine throne and sometimes wore a mural crown (like a Tyche). On the other hand, Allat—o Al Uzza in Petra—was Dushara's main partner. As Al Uzza is often equated with Aphrodite/Venus and there are no iconographic elements at Hurawa to associate the goddess there with Venus, then it is possible that the broken cult statue of the goddess represents Allat. The temple would then be dedicated to the supreme Nabataean couple Dushara-Allat so frequent in Nabataea outside Petra proper.

The 'Zodiac' is not the only component of the temple decorative programme with a strong astral symbolism (this will be analyzed in details further below). Huge sections of the decoration of the temple also have such a symbolic nature. On the façade, there were seven unframed busts of the celestial deities, representing the visible planets, of which only Helios with a radiate crown, Kronos-Saturn with a reaping hook and Hermes-Mercury with a cap have been well preserved. The portico was framed by the main deities of the temple: a Storm-God with thunderbolt and a Tyche. And in the corners there were the unframed busts of Apollo and a male deity with a crescent Moon behind that are difficult to interpret, perhaps a form of the Semitic lunar god Sin. To further complicate the situation, a female statue of a goddess with attributes of the iconography of the Greek sun god Helios has been found on the site. Does this reflect the female nature of the Sun deity in ancient Arabic culture? Finally, Isis is also represented in a small cult statue.

All in all, given the celestial iconography programme, with personifications of both the visible planets and zodiacal signs, not only in the Zodiac but also in the Period 3 frames, the presence of worshippers overnight at Khirbet et-Tannur, perhaps to watch the night sky or await the rising Sun might be expected, along with night-time rituals, as archaeology seems to prove (McKenzie 2013). Indeed, the hilltop location would have provided a clear and excellent view of astronomical events.

But Hurawa, if this is the correct name for the site, was not isolated. A few kilometres south stands the ruins of Khirbet edh-Dharih, a small village on the eastern bank of the Wadi Al-La'aban which was possibly known as Aïn La'aban in Antiquity thanks to the presence of a permanent spring in the area (see Fig. 8.2). Thamudic inscriptions found on the site and dated ca. CE 32/36 mention the *Bayt li*

Lat in relation to the site. This is most probably a reference to a shrine in the area that was converted into an impressive temple (Fig. 8.2, panel b) almost certainly at the same time that the building of Period 2 occurred at Hurawa due to the extreme similarities of the decorative programme (with the representation of the planets and personification of the zodiacal signs) and the building techniques.

Given the contemporaneity of their construction, and the fact that the busts on the exterior of one depict the planets and on the other the zodiac, with the reverse inside (i.e., in Dharih the personified zodiacal signs are in the exterior façade) this also suggests a relationship between the two temples. Actually, it is highly probable that people normally residing in Aïn Al-La'aban were the same ones who handled the cult in Hurawa at certain key times of the year when pilgrimages and festivals would connect the two sites through the *Via Nova Traiana* (see Fig. 8.2). The hill-top sanctuary could not have permanently housed a great number of people and the daily cult could have been easily handled by a reduced number of priests. The earthquake of CE 363, which destroyed Hurawa, also caused much destruction in Khirbet edh-Dharih but activities on site endured so that it was not deserted until the sixth century CE.

The pilgrimages between the two sites, and the corresponding festivals in Hurawa, were probably performed on certain key dates of the new calendar of the *Provincia Arabia* which was inaugurated in CE March 106, retrospectively. These possibly were a reflection of previous important festival dates of the Nabataean lunisolar calendar. Worshippers could have come at specific times of the week, month or year to make their offerings and celebrate meals together. When were those festivals? To further analyze this possibility we must first discuss the most challenging evidence found in Hurawa, the so-called 'Zodiac'.

Khirbet-et Tannur's New Paradigm: Zodiac vs. Almanac

Figure 8.3 shows a close-up of a reproduction of the stela were the two larger fragments have been glued together to show the complete aspect that the piece would have had on its cult site in the temple. The centre of the representation is dominated by the image of a goddess with a mural crown over her head (a sort of Tyche). She is accompanied by a crescent Moon and a unique object behind her shoulder: two converging sticks, one tipped by an ear of wheat or a small pine cone, and the other similar to a pointed crescent Moon. This symbolism has been found in other representations at Tannnur, so she is also the 'Tyche' of Khirbet et-Tannur and thus perhaps another manifestation of Allat. Since the discovery of this masterpiece upper fragment in Glueck's excavation and the serendipitous finding of the lower part of the 'Zodiac' (with the two lowermost signs and the holding Nike frieze) there have been several attempts to interpret the strange distribution of the zodiacal signs divided into two semi-circular sections, one to the left for the spring and summer signs, starting by a personified Aries, and one to the right for the autumn and winter signs, staring by a personified Libra, with the scales to avoid confusion (see Fig. 8.3).

Glueck himself (1952) was unable to offer a satisfactory explanation for this strange distribution apart from a mere reasoning that it would divide the year into two halves starting by the spring and Spring Equinoxes, respectively. He could not find an explanation either for the dots appearing in the signs of Aries (two of them), Gemini, Leo and Capricorn. So he asked for help from one of the best historians of astronomy of the epoch, the Harvard Professor Owen Gingerich who followed a logical pattern. If the image ought to be interpreted as a zodiac, then the dots could be a representation of the position of the five visible planets (excluding the two luminaries) and the whole diagram read as an astrological chart. Gingerich (1966) cleverly thought of an epoch close to the Spring Equinox with the Sun in Aries which would locate Mercury and Venus nearby. He found an apparent solution for 22 March 82 CE with Mercury and Venus located in Aries, precisely, Mars in Gemini and Jupiter in Leo. Saturn would be in Pisces which had not been preserved. This solution left one dot in the personified Capricorn (a young man) without an explanation, but was the best that could be found. When the lowest part of the diagram was found in the 1950s, Gingerich's hypothesis was weakened when a dot was absent in Pisces. Our own analysis, with the most powerful modern celestial charting techniques and programs, has been unable to find any other reasonable alternative for an astrological explanation covering the three centuries that the temple was in use, and even before. To our mind, the diagram is not a sort of horoscope and perhaps not even a zodiac.

Dozens of different hypotheses have been developed to explain what this diagram represents, most of them being sad examples of *vox nihili*. For completeness, we may discuss the most recent of them. Rosenthal-Heginbottom (2001) made an attempt to understand the significance of the 'Zodiac' within Nabataean religion and art. She identified Aries as a female bust that might be identified with Athena. She argued that the distribution was by no means accidental, and that it was the result of an astrological concept and astronomical knowledge, even though its practical and spiritual significance was not fully understood. The peculiar halving was again explained as a calendar year with two beginnings: the civic year in autumn and the religious year in spring. On the contrary, Janif (2006–2007: 348) argued that "... the order of the zodiac constellations at Tannur would therefore respond to some theological and even gnostic concerns, rather than to a method of time reckoning among the Nabataeans ...", but he did not give a clue as to these hypothetical concerns. One of the most reputed scholars on Nabataean religion, Robert Wenning (2009), tried to find a solution fueling his expectancies: in Babylonian mythology the Ram is connected with the Moon, the Scales with the Sun, and both the Moon and the Sun are male deities. If this diagram has a Middle East root, this might explain why Aries is depicted here as a male bust—notice the contradiction. Virgo and Pisces are related directly to the Earth itself and its fertility. Hence, the composition does not follow two calendars but it rather reflects a 'cosmological programme' (sic).

Finally, Judith McKenzie, the person who knows better this fantastic artwork, makes a reasonable assumption: the division of the Zodiac and the orientation of the sanctuary—'facing east precisely'—accord with a major spring festival at Khirbet et-Tannur in Nisan to mark the New Year. Such a festival occurred there in the

Vernal Equinox or in the first New Moon after this. Perhaps there also was a harvest festival on the Autumnal Equinox (McKenzie 2013; McKenzie et al. 2012). However, "... the significance of the dots remains a mystery ..." for her (McKenzie et al. 2012: 394) a very wise cautious position.

This was the situation when we decided to change our minds completely in order to find a solution for the peculiar zodiacal diagram and its special characteristics, including the presence of the elusive dots. Two considerations should be taken into account. On the one hand, that the Nabataeans had used the Babylonian lunisolar calendar (with a whole set of festivals of presumably astronomical characteristics; see earlier sections) until the annexation of the kingdom and the creation of the Roman *Provincia Arabia* in 106 CE, which established officially a new dating system in accordance to the Julian calendar. This would imply the need for a completely different way of timing the epoch of traditional festivals and pilgrimages within a completely new chronological framework. The similarity between the new calendar selected (12 months of 30 days plus five additional days) and the Egyptian time keeping system (Belmonte 2003, 2012) would also be inspiring.

On the other hand, we recall the attention to the way in which the Greco-Roman world used to mark specially signaled days or feast in their calendars—a tradition that in certain aspects has survived to the present day in our home and work almanacs—but that in antiquity took a clear, on occasions monumental, methodology: the *parapegmae*. In these devices, special small holes or plugs were distributed all along the different kind of time markers (days of month, months, weeks, etc.) where wood or metals sticks could be positioned to appropriately mark the timing of special events. One of the most famous, and most relevant for our discussion, was the one found at the Baths of Trajan in ancient Rome (see Fig. 8.4).

This parapegma (Bianco 2017) includes the signs of the zodiac and the faces of the planetary deities (Saturn and Jupiter were erased for unknown reasons but can be reconstructed, see Fig. 8.4) in a complete parallelism to the decorative programme of the temples at Aïn Al-La'aban and certainly Hurawa where our 'Zodiac' was found. The tablet was found in Trajan's Baths in Rome (i.e. the Emperor who annexed Nabataea), which were constructed in CE 104–109 and operated well into the early fifth century. It cannot be determined exactly when this parapegma was used but the presence of the seven planetary deities representing the 7 days of the week perhaps indicates that this masterpiece ought to have a later date within that time interval. It features pegs that were moved from day to day and month to month. Interestingly, the zodiacal signs are represented instead, or in substitution, for the corresponding months, suggesting that in the ancient Roman world, zodiacal signs could have been considered as alternative iconographic representations for the months of the year, which had no proper standardized symbolism for themselves. Interestingly, only the days of the month up to the 30th could be marked in the tablet. Either the months with 31 days operate differently or we are facing a strange arrangement of a solar year, but not necessarily of the Julian calendar.

What is really important for our purposes is that the zodiacal signs were not in representation by themselves but were instead an alternative iconography for the months of the year. Could the diagram found at Khirbet et-Tannur be a peculiar type

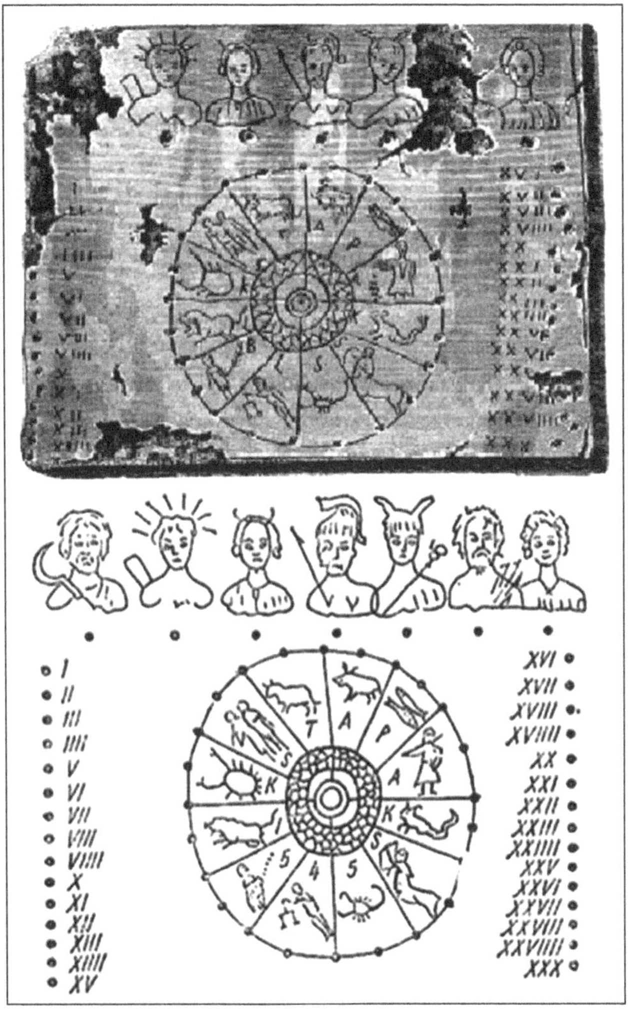

Fig. 8.4 The Parapegma table found at Trajan's Baths in Rome. The lower panel shows a reconstruction of this interesting piece where the broken sections have been artistically restored. This device was able to use small pegs to indicate the day of the week, the month (position at either *Calendas* or *Idus*) of a Julian-type calendar, through the equivalent zodiacal signs, and the 30 days of a month (diagram by the authors, on an image by courtesy of the Museo della Civiltà Romana)

of parapegma with a similar symbolism? This would mean that the zodiacal signs will not be there representing themselves—and hence the piece would not be a 'Zodiac'—but would represent the months of the year and hence the diagram is actually a calendar or almanac or, as we shall see, a parapegma in its own right. Which calendar, and months, could be imagined here? Certainly, if Period 2 of Hurawa corresponds to the first half of the second century CE and the decoration

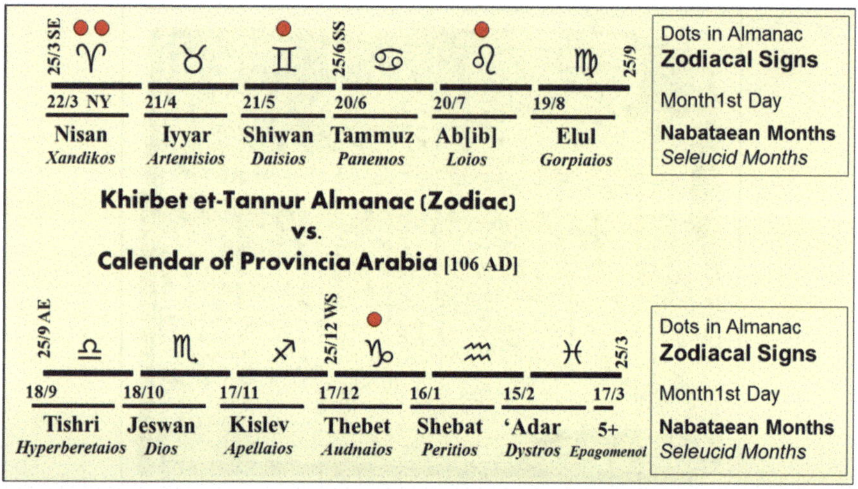

Fig. 8.5 A diagram schematically representing the parallelisms between the Khirbet et-Tannur 'Zodiac' and the months of the Calendar of the Roman province of Arabia, inherited from the months of the Nabataean lunisolar calendar of Seleucid inspiration. The dates of the solar time-markers (solstices and equinoxes) and of the beginning of the months are specified. The small circles marked the same position of the dots at the 'Zodiac' itself. See the text for the corresponding discussion (diagram by the authors)

programme of the main altar is contemporary with that building phase, then the calendar represented must be that of *Provincia Arabia*, inaugurated in 22 March 106 CE—which was also a New Moon and hence the beginning of the previous lunisolar year, now to be abandoned, at a date nearly simultaneous to the annexation of the Nabataean Kingdom by Emperor Trajan.

Figure 8.5 shows a schematic diagram of the Tannur Almanac with its zodiacal signs in comparison to a complete year of the calendar of the *Provincia Arabia*. This calendar operated as the Julian calendar—with a bissextile after every 4 years—but with an Egyptian structure of 12 months of 30 days plus the 5 epagomenal days at the end of the year. The name of the months were those of the previous Nabataean lunisolar calendar (and their Seleucid alternatives on Greek inscriptions, see Fig. 8.5) which order was respected, but that will not follow the lunar cycle anymore, and their Julian dates counterparts. It started on Nisan or Xandikos 1st, roughly corresponding to 22 March in the Julian calendar. It is worth mentioning that the dates of the vernal and Autumnal Equinoxes and of the Summer and Winter Solstices had been fixed at 25 March, 25 September, 25 June and 25 December in the Julian calendar, respectively (González-García and Belmonte 2006).

One thing immediately caught our attention. The dots within the Tannur Almanac are located within those months which are most frequently mentioned in Nabataean monumental inscriptions (Belmonte and González-García 2017) and that in the introduction were identified as those months where the most important festivals—with minor exceptions—ought to be timed. Figure 8.6, illustrating this fact, is included for

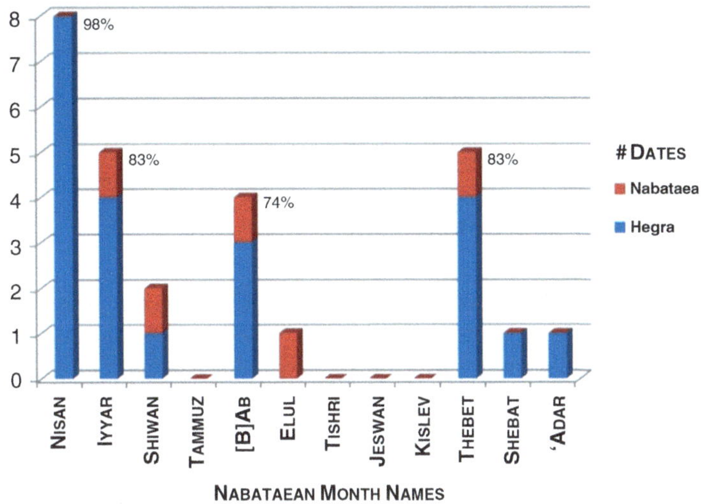

Fig. 8.6 Histogram of the months of the year mentioned in dated Nabataean inscriptions in the tombs of Hegra and elsewhere in Nabataea, including Petra, during the Kingdom period. Notice the relevance of the months of Nisan, Iyyar, (B)ab or Ab'ib and Thebet where the number of mentions are statistically significant (numbers indicate the probability of the month not being mentioned by chance). See the text for further discussions (adapted from Belmonte and González-García 2017)

direct comparison. The cases of Shiwan, Ab[ib] and Thebet are clear. However, there are two dots in Aries. This sign could stand for Nisan but also for part of the month Iyyar. The parallelism is certainly striking.

All in all, if each of these considerations is taken into account, one implication becomes apparent: the dots within the Tannur Almanac could be playing the role of holes and plugs in a standard parapegma. In this particular case, our hypothesis is that these five dots would be marking five most relevant festivals of the old lunisolar calendar to be celebrated in Hurawa, but now conceived within the framework of the new calendar of the Province of Arabia and its New Era. These might be, according to our previous classification: (1) the Full-Moon of Nisan, equivalent to Hebrew Passover; (2) the First Crescent Moon of Nisan or of Iyyar, first one of the yearly calendar depending of which year is considered; (3) the First Crescent of the lunar month including the Summer Solstice, or the Summer Solstice itself; (4) the heliacal rising of Sirius related to Isis-Al Uzza—remember that the calendar has a clear Egyptian inspiration; and (5) Dushara's Winter Solstice festival in dates close to the Epiphany. All these dates could be related to relevant festivals or pilgrimages associated with the divine couple revered on the temple of Khirbet et-Tannur, coming either from Aïn Al-La'aban or elsewhere in ancient Nabataea, now the province of Arabia.

To these, two additional feasts could be associated with New Years's Eve in Nisan 1st, and an additional festival at the beginning of Tishri which could be somehow related to the month of Aggathalbaeith. These might explain the double nature of the almanac divided into two clear halves. It is indeed the moment to change our mind and develop a new paradigm in which the 'Zodiac' of Khirbet et-Tannur should be better known as the Tannur Almanac or, even better, as the Tannur Parapegma.

Final Discussion and Conclusion

Khirbet et-Tannur (Hurawa) may have been a pilgrimage centre, possibly even a national shrine, during most of its history. This was especially the case during the phase when the splendid temple, with a suggestive astral decorative programme, was in use. This mostly coincided with the Era of *Provincia Arabia*, from its inception in CE 106 through to the earthquake in CE 363 when the place was destroyed, burnt and abandoned. The almanac in the main altar of the temple served as an attractive souvenir of the activities and festivals—and their timing—that ought to be celebrated in the sacred shrine. But this was not all. The temple orientation gives a further clue.

Our investigation of the site (Belmonte et al. 2013) produced an azimuth of 92½° with an angular height of the horizon in that direction of c. 2° (the temple is not topographically orientated to Djebel Tannur which lies further to the northeast; see Fig. 8.2). Interestingly, this gives a value for the declination (δ) of about −1¼°, corresponding to a date 3 days before the Spring Equinox (Fig. 8.2, panel d). This is exactly (with the margin of 1 day depending on the leap nature of the year) 22 March, i.e. Nisan first of the province of *Arabia* calendar. Hence the temple is orientated towards the sunrise at New Year's Eve in the New Era! This fact alone would justify most of the arguments that have been analyzed and discussed in the course of this paper, including the pilgrimages, overnights on site and the need for open skies at the mountain summit, among many others. The rest of the feast to be celebrated on site would be marked by the dots in the appropriate positions on the almanac to be found in the sancta sanctorum of the temple.

Additionally, an orientation 3 days before the Spring Equinox also reflects a pointing to sunrise 3 days after the Autumnal Equinox. Figure 8.5 clearly illustrates that this would mean a date close the Tishri 10th.[3] We could recall Epiphanius informing us that his contemporary Arab tribes of southern Palestine and Transjordan performed a pilgrimage during the month of Aggathalbaeith to a major sanctuary in the region (Panarion 51, 24). The time of the pilgrimage would have corresponded to the month of Tishri in the new calendar of Arabia. Was Khirbet et-Tannur the Baeith (of God), the focus of this pilgrimage? While the parallelism is suggestive, we do not have all the clues to arrive at a positive answer. It could easily have been a sanctuary in Petra, or elsewhere in the area, instead.

[3]The hajj in the present day lunar Islamic calendar starts of the 10th of Dhu al-Hijja, a month that could be paralleled with Tishri before intercalation was forbidden in the Quran (9, 37).

Interestingly, on our last visit to Petra we learnt that recent excavations at Petra have found the suggestive remains of a small Imperial cult temple beside Qsar al-Bint. It was apparently erected by CE 106–114 and has the same orientation as the city Cardus or colonnade street. This has an orientation of 97½° directed towards a very high eastern horizon with an angular height of c.10°. These numbers again render a declination of −1¼°, exactly the same than in Hurawa. Is this a coincidence? We do not think so. The Romanization process of the Nabataean capital is still poorly understood and it is highly probable that, in this jewel of their ingenuity, the Nabataeans' ability to adapt their ancient customs to the new necessities and obligations was fully into operation.

The temple at Hurawa, its orientation and its fascinating decorative programme, together with Petra itself, could simply be the two effective faces of a same coin: to adapt or die!

Acknowledgments This work has been financed under the framework of the projects P/310793 'Arqueoastronomía' of the IAC, and AYA2015-66787 'Orientatio ad Sidera IV' of the Spanish MINECO.

References

Alpass, P. (2010). The Basileion of Isis and the religious art of Nabataean Petra. *Syria, 87*, 93–113.

Al-Salameen, Z., & Falahat, H. (2009). Religious practices and beliefs in Wadi Mousa between the late 19th and early 20th centuries. *Al Majaq al Urdunia, 3*, 170–204.

Alzoubi, M. (2016). The Nabataean timing system. *Acta Orientalia, 69*, 301–309.

Belmonte, J. A. (1999). Mediterranean archaeoastronomy & archaeo-topography: Nabataean Petra. In A. Lebeuf & M. Ziolkowsky (Eds.), *Proceedings of the Vth SEAC conference* (pp. 77–90). Warzaw: Institute of Archaeology of the University of Warzaw.

Belmonte, J. A. (2003). Some open questions on the Egyptian calendar: An astronomer's view. *Trabajos de Egiptología (Papers on Ancient Egypt), 2*, 7–56.

Belmonte, J. A. (2012). *Pirámides, Temploa y Estrellas: Astronomía y Arqueología een el Egipto Antiguo*. Barcelona: Crítica.

Belmonte, J. A., & González-García, A. C. (2017). Petra re-visited: New clues for the Nabataean cultic calendar. *Culture and Cosmos, 21*, 131–150.

Belmonte, J. A., González García, A. C., & Polcaro, A. (2013). Unveiling Petra: Astronomy and landscape in the Nabataean lands. *Nexus, 15*, 497–502.

Bianco, B. (2017). *What's a parapegma?* https://dayology.com/2017/03/24/the-trajans-baths-parapegma/

Browning, I. (1989). *Petra*. London: Chatto and Windus.

El Khoury, L. (2006–2007). Nabatean pilgrimages as seen through their archaeological remains. *ARAM Periodical, 18–19*, 325–340.

Forcada Nogués, M. (1993). *Tratado Sobre los Anwa y los Tiempos de Ibn 'Asim* (Fuentes Arábico-Hispanas 15). Barcelona: C.S.I.C.

Gawlikowski, M. (1990). Les dieux des Nabatéens. *Aufstieg und Niedergang der römischen Welt II, 18*(4), 2659–2677.

Gingerich, O. (1966). A note on attempts to date the Zodiac. In J. McKenzie (Ed.), *The Nabataean temple at Khirbet et-Tannur, Jordan: Final report on Nelson Glueck's 1937 excavation* (p. 228, Appendix 3.1; The Annual of the American School of Oriental Research 67, 2013). Oxford: Oxford University Press.

Glueck, N. (1952). The Zodiac of Khirbet et-Tannur. *Bulletin of the American Schools of Oriental Research, 126*, 5–10.

González-García, A. C., & Belmonte, J. A. (2006). Which equinox? *Archaeoastronomy. Journal for Astronomy in Culture, 20*, 95–105.

Hawting, G. R. (1999). *The idea of idolatry and the emergence of Islam. From plemic to history.* Cambridge: Cambridge University Press.

Healey, J. F. (1993). *The Nabataean tomb inscriptions of Mada'in Salih.* Journal of Semitic Studies, Supplement 1. Oxford: Oxford University Press.

Healey, J. F. (2001). *The religion of the Nabataeans: A conspectus.* Leiden: Brill.

Hoyland, R. G. (2001). *Arabia and the Arabs: From the Bronze Age to the coming of Islam.* Oxon: Routledge.

Janif, M. M. (2006–2007). Sacred time in Petra and Nabataea: Some perspectives. *ARAM Periodical, 18–19*, 341–361.

McKenzie, J. (2013). *The Nabataean temple at Khirbet et-Tannur, Jordan: Final report on Nelson Glueck's 1937 excavation.* The Annual of the American School of Oriental Research 67. Oxford: University of Oxford Press.

McKenzie, J. S., Reyes, A. T., & Greene, J. A. (2012). The context of the Khirbet et-Tannur Zodiac, Jordan. *ARAM Periodical, 24*, 379–420.

Merklein, H., & Gerger, Y. (2001). The veneration place of Isis at Wadi as-Siyyagh, Petra: New research. *Studies in the History and Archaeology of Jordan, 7*, 421–432.

Mordtmann, J. H. (1875). Dusares bei Epiphanius. *Zeitschrift der Deutschen Morgenländischen Gesellschaft, 29*, 99–106.

Rodríguez-Antón, A., Belmonte, J. A., & González-García, A. C. (2016). Romans in the Near East: Orientation of Roman towns and forts in modern Jordan. *Mediterranean Archaeology and Archaeometry, 16*, 153–160.

Rosenthal-Heginbottom, R. (2001). A Nabataean Zodiac. In *Measuring and weighing in ancient times* (pp. 51–54). Haifa: Reuben and Edith Hecht Museum.

Samuel, A. E. (1972). *Greek and Roman chronology: Calendar and years in classical antiquity.* Handbuch der Altertumswissenschaft, 1. Abt. 7. Munich: Beck.

Stern, S. (2012). *Calendars in antiquity.* Oxford: Oxford University Press.

Wenning, R. (2001). The betyls of Petra. *Bulletin of the American Schools of Oriental Research, 324*, 79–95.

Wenning, R. (2009). The message of the Khirbet at-Tannur reliefs. *Studies in the History and Archaeology of Jordan, 19*, 577–584.

Chapter 9
The Archaeoastronomy and Chronology of the Temple of Jupiter at Baalbek

Giulio Magli

Abstract The temple of Jupiter at Baalbek in Lebanon is one of the most complex architectural projects ever conceived. Several issues remain unsolved about this site; in particular, the relative chronology and the dates of construction of the two 'podia' of the temple are unsure, as well as the true nature of the cult of Jupiter practiced there. We present here a new architectural analysis based on the orientation and on other features of the temple, which clearly point to a unified project originally conceived under Herod the Great.

Introduction

The Temple of Jupiter at Baalbek (Heliopolis), Lebanon, is famous worldwide for its megalithic architecture (Segal 2013). Yet an impressive number of problems remain unsolved about this monument, including its precise dating, the phases of construction, and even the true nature of the 'triad' cult involving Jupiter, Venus and Mercury that is known to have been practiced at this place (Kropp 2009, 2010). From the point of view of the history of architecture, one of the problems is the absence of contemporary sources, but another is the uniqueness of the megalithic building technique.

The main features of the temple can be briefly described as follows (Fig. 9.1). The complex develops along a monumental axis comprising a hexagonal court and a huge platform with Propilea, both built by the Romans in the second century AD, together with the nearby temples of Bacchus and Venus. The Jupiter Temple proper is located at the end of the Propilea. Its final phase of construction—with the erection of the enormous columns almost 20 m high—is dated to the Julio-Claudian period (AD 40–60) due to a graffito left by one of the stonemasons, which mentions the date 2 August 60 AD. This was found on top of one of the column shafts, and construction therefore must have been almost finished by that date. The columns rest on a

G. Magli (✉)
Department of Mathematics, Politecnico di Milano, Milan, Italy
e-mail: giulio.magli@polimi.it

G. Magli et al. (eds.), *Archaeoastronomy in the Roman World*, Historical & Cultural Astronomy, https://doi.org/10.1007/978-3-319-97007-3_9

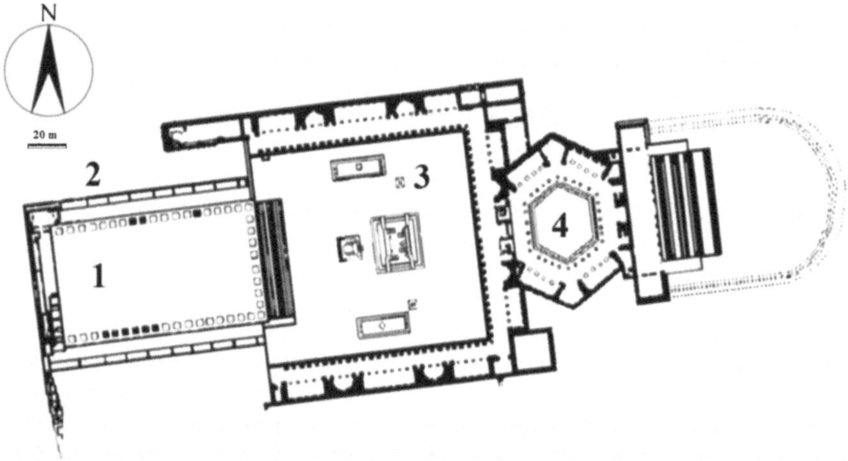

Fig. 9.1 Plan of the Temple of Jupiter at Baalbeck. (1) Podium I, (2) Podium II, (3) Propylea, (4) Hexagonal court

huge basement which, adopting the terminology of Kropp and Lohmann (2011), will be called Podium I.

At a distance of *a few meters* from Podium I runs a huge wall, 'Podium II', which is parallel to Podium I and surrounds it along three sides forming a giant U-shaped structure. This structure was originally built without any connection to Podium I whatsoever, as is apparent upon looking at its north-west side. The design is astonishing: it is constructed through the superposition of increasingly greater stones as the height increases. Big stones are in fact used at the base, but even larger stones are present in the second course, and huge megaliths (about 500 tons each) were used to build the third course. Finally, enormous blocks, each around $4 \times 4 \times 20$ m and not less than 800 tons, were to be placed in the uppermost course, but only the south-west side was completed, where the three famous stones that are usually called (somewhat inappropriately) the 'trilithon', were put in place. At least three other enormous blocks remain at the quarry some hundreds of meters to the south-west, including the largest of all, which has only recently been discovered (Abdul Massih 2015).

This construction technique is very strange indeed, because the use of pre-compression in megalithic masonry—namely, putting in place huger stone blocks near the summit instead of the base—is very well attested in antiquity but in polygonal courses, where it was used to 'frieze' stresses and therefore to strengthen the joints between the blocks lying beneath (see e.g. Magli 2006). Clearly such a trick does not work if the stones are placed on horizontal layers, as at Baalbek, so—although it may seem incredible—we have to conclude that only aesthetic reasons inspired the builders.

Who built the two podia, and when? The most accepted archaeological viewpoint was that the building belongs to the Julio-Claudian period, with perhaps two phases of construction. However, a recent architectural analysis of Podium I has shown striking similarities with Herodian sanctuaries, such as the use of alternating rows of headers and stretchers and the presence of drafted-margin masonry. In particular, obvious similarities exist with the Herodian phase at the Temple Mount at Jerusalem, not only in general appearance, but also in proportions and measures. All in all, this new analysis leads quite naturally to the conclusion that Podium I was originally built by Herodian architects (Kropp and Lohmann 2011; Lohmann 2008, 2010). In this connection it should be noticed that Baalbek was not enclosed in Herod's reign; however, it was enclosed in the Roman colony of Berytus (Beirut), founded in 15 BC. Since this date, Herod is known to have been keen in showing his attention to the Roman possessions in the whole area.

So far so good for the date of the original construction of Podium I. There remains, however, the problem of dating Podium II which, as mentioned, is structurally unrelated to Podium I. According to Kropp and Lohmann (2011), it postdates the Herodian phase and was designed by the Roman architects to 'harmonize' the dimensions to Roman standards. The idea that the Roman builders, for whatever reason, decided to carry on a megalithic enterprise that had no antecedents whatsoever in the whole history of architecture, not to say the Roman one, is however frankly difficult to believe, especially if we take into account the well-known practical mentality of Roman engineers. It is thus the aim of the present paper to re-analyze the problem of the date of construction of the Temple of Jupiter. We will start from the point of view of modern archaeoastronomy (see e.g. Magli 2015), and therefore we will study the building within the sky landscape in which it was immersed, in strict connection with the historical context. As we shall see, our results will finally lead to a completely new hypothesis for the dating of Podium I, that supports the idea that both Podia were originally conceived together, within the context of a single-phase Herodian project.

Archaeoastronomy at Baalbek

Orientations of temples in the cultures of the Mediterranean have been widely investigated (see e.g. Boutsikas 2009; Belmonte et al. 2009) but to the best of our knowledge the orientation of the Baalbek temple has never been studied.

The area is quite well covered by satellite imagery and as a consequence from satellite data (Google Earth and Bing) both the azimuth and the horizon height (from inside looking out) can be determined with good approximation (say to within 0.5° for both). The results are azimuth 75° 30′, horizon height 5°. Using the program getdec (kindly provided by Clive Ruggles) which takes into account refraction, at the latitude of Baalbek these data yield a declination of ~14° 44′. This declination is within the solar range: the Sun therefore rises in alignment with the temple twice a year, around 1 May and 12 August (Gregorian; up to the second century AD the difference with the Julian was negligible). These dates of course are not of special significance for the solar cycle, and this may be seen as a confirmation of already existing doubts (Kropp 2010) about the true solar character of the 'Heliopolitan' (*sic*)

Jupiter. The dates are not close enough to days of special significance in the Roman calendar either. For instance, the foundation of Rome on 21 April is known to be referenced in Roman architecture (Hannah and Magli 2011) but the movement of the rising Sun along the horizon at the end of April is too fast to think this could be an orientational error. Also, these dates do not match anything significant in the Hebrew luni-solar calendar, as the closest important event is Passover which however never extends to the end of April.

The orientation does not match any recognizable pattern for comparable temples either. In fact, the orientations of the other three main temples of Zeus in the region, namely Kanawat, Damascus, and Gerash, are as follows. The Kanawat temple points almost to true north (azimuth 4°, horizon flat or nearly flat). Interestingly, the building is very clearly directed to the center of Philippopolis, which lies some 10 km to the north and was probably planned a few decades later, using the temple as a landmark. The temple of Damascus (azimuth 85°, horizon nearly flat) has a generic solar orientation which however conforms to the orthogonal grid of the town. Finally, the huge sanctuary of Zeus at Gerash has been measured on site by González-García et al. (2016). According to their data (azimuth 54°, horizon height 3° 30′, declination +31° 30′) the temple is out of the solar range and not far from the maximal northern standstill of the Moon, a possibility that certainly deserves further research but which in any case does not match Baalbek. Last but not least, the Baalbek azimuth is not governed by the topography either—the huge platform was oriented exactly where the builders wanted it to be, without any geological constraints.

If we exclude chance, the only remaining possibility is a stellar alignment. To investigate this possibility we start by analyzing the sky at Herod's time. It is then seen that a rather important celestial object was rising in alignment with the temple: the Pleiades. Of course the Pleiades are an asterism, not a single star (seven stars can be distinguished with the naked eye); however, they can be considered (and were considered in antiquity, since the time of Hesiod in the Greek world) as a single entity spanning ~0.5° in declination. Their declination in Herod times was between 15° 30′ and 16° (for instance, the star Alcyone in 15 BC had a declination of 15° 50′). The agreement with the temple declination is therefore good, and the horizon height which corresponds to the temple front assures that the asterism was really visible (stars are not visible until they are at a height at least comparable to their magnitude in degrees).

Interest for the Pleiades is well documented in Greek religion; for instance, the role of this asterism has been shown to be fundamental to the rites at the Artemis Orthia sanctuary in Sparta (Boutsikas and Ruggles 2011; see also Boutsikas and Hannah 2012 for the role of the Hyades at Athens' Acropolis). Is it possible to associate the Pleiades with the Heliopolitan Jupiter? The pre-Roman history of the God is uncertain, but the iconography is well attested from the Roman period and from the unique written source we have about Baalbek, the Saturnalia dialogues of the fifth-century author Macrobius. The cult image of Jupiter represented a young, unbearded God, bringing a huge vase-shaped top hat (Kalathos). The God usually brings also grain ears and a whip, and is accompanied by two walking bulls. The

Heliopolitan Jupiter thus had the clear attributes of a God of fertility. In this connection, the orientation to the Pleiades becomes more understandable. Indeed already in the Hesiod calendar (eighth century BC) the harvest time of the cereals was indicated by the heliacal rising of Pleiades which occurred in the first week of May. Taking into account the shift in time but also that in latitude, this date does not change substantially in Baalbek. Indeed, we can estimate the date of the heliacal rising at the end of the first century BC in Baalbek using planetarium software (here we use Starry Night Pro 8) and considering that a star of magnitude M starts to be visible when it is separated from the Sun by about 4 M°, that is, the star is at M degrees when the Sun is under the horizon at least 3 M°. Assuming the magnitude of the Pleiades—considered as a single object—to be around 1.6, we can see that these conditions are satisfied around 5 May. Interestingly enough, therefore, the alignment of the temple approximates the direction of the heliacal rising of the Pleiades, and that of the rising Sun a few minutes later, on the same days: a quite peculiar coincidence.[1] In this connection it may further be noted that beforehand on the same nights, one could see in the very same direction the rising of the constellation Aries (the declination of the star Hamal was ~13°) a constellation associated with spring and renewal as well (a direct association of Aries with Zeus also probably existed at those times, but it is securely documented only in the case of Zeus-Ammon, the horned God of Egyptian origin which is not present at Baalbek).

Discussion and Concluding Remarks

As a matter of fact, the Pleiades appears to be the unique feasible explanation for the orientation of Baalbek's Temple of Jupiter. Due to the precession of the Earth's axis however, the declination of each star slowly and continuously changes. In the case of the Pleiades, the alignment at Baalbek worsened slight with time: for instance in AD 60 Alcyone had a declination of 16° 15′. A corresponding slight deviation of the axes of Podium II with respect to Podium I might thus have provided a clue to its dating, but this is not the case. Thus, we are led to think either that the builders of Podium II were interested only in the solar alignment for reasons we do not know, or a new possibility: contemporaneity of construction of the two Podia. To show the feasibility of this latter hypothesis we will proceed *ab absurdo* by showing the weakness of the other two possibilities.

The first hypothesis (which prevails in non-scholarly publications) is that Podium II predates Podium I. It is easy to see, however, that any sensible architect willing to build Podium I after Podium II would have used it. Only a fool would construct

[1]It should, however, be noted that the dates of Heliacal phenomena are always difficult to identify in a precise way. In particular, a more prudential estimate (by increasing the distance in height from the Sun, or by decreasing the assumed magnitude of the asterism, or both) would lead to shift the date later in May.

ex-novo a huge basement, oriented in the same way and accurately placed just a few meters inside the existing wall, without taking advantage of it as a ready-to-go, tremendously stable and affordable structure.

This observation, at least in the author's view, clearly dispatches the "Podium II predates Podium I" theory. Also the second possibility, that Podium I predates Podium II, is quite problematic. As mentioned, it implies that the style of Podium I was not acceptable to the Roman standards and therefore in Julio-Claudian times the Romans opted for the enlargement of the building (Kropp and Lohmann 2011). However, again, any sensible architect willing—for 'stylistic' reasons that, at least to the present author, seem quite weak on their own—to enlarge Podium I up to the dimensions of Podium II, would have used the pre-existing structure and expanded the basement up to the desired dimensions. Constructing ex-novo a self-standing, gargantuan megalithic wall is almost as illogical as the one implied by the inverse chronology.[2] Accordingly, I propose here that the absence of structural connection, and simultaneously the strict parallelism, between the two Podia can be explained much better if the structures were planned together (but a possible explanation for the fact that they were not constructed as a connected building will be given below).

In accordance with the Herodian dating proposed by Kropp and Lohmann for Podium I, I propose that the whole project was conceived under Herod the Great. In this respect it should be noted that strict architectural analogies with the Herodian architecture at the Temple Mount do hold also for Podium II. In fact a wall made of gigantic stone blocks has been unearthed in the tunnels along the western side of the Mount (Bahat 1994; Ritmeyer 1992). These blocks show beyond any reasonable doubt that megalithic masonry was in the minds, and within the abilities, of Herodian stonemasons: the largest known of the Herodian blocks in Jerusalem is indeed 13.7 × 3.2 × (probably)3 m, and weighs about 570 tons. Furthermore, the wall in itself is very similar to that of Baalbek's Podium II; for instance, the hugest stones are set over courses of smaller blocks.

Why did Herod's architects built Podium II as a disconnected unit? A possibility is that they wanted to form a U-shaped gallery encircling the sides of the temple. The function of the back gallery might have been related to the cult, perhaps to exploit oracular rites. The gargantuan project remained unfinished and, in particular, the builders did not succeed in completing the exterior side walls transporting the missing megaliths, so the construction of the vaults did not begin. As a consequence, the megalithic wall remained as a sort of (at this point really anti-esthetic) curtain and this explains why at the Julio-Claudian stage it was decided to fill it with blocks of

[2]The presence of Roman sketch engravings of the temple pediment on one of the blocks of the Trilithon has been claimed as a proof of contemporaneity. Of course it is not: the Roman architects used to sketch their projects on pre-existing monuments, for instance on the paved floor in front of Augustus mausoleum a precise drawing of Hadrian's Pantheon pediment can be seen. Another proof should be that in the lower course of Podium II a piece of a column drum was used instead of a block; however—if it was not a Arab repair—the small piece does not appear to belong to the huge columns of the Julio-Claudian phase and may come from the Herodian temple.

stone. The temple we can see today is the final result of the Arab conversion of the building in a fortress, with walls built with second-use blocks.

References

Abdul Massih, J. (2015). The megalithic quarry of Baalbek: Sector III the megaliths of H. ajjar al-H. Ibla. *Journal of Eastern Mediterranean Archaeology and Heritage Studies, 3*(4), 313–329.

Bahat, D. (1994). The Western Wall tunnels. In H. Geva (Ed.), *Ancient Jerusalem revealed* (pp. 97–118). Jerusalem: Israel Exploration Society.

Belmonte, J. A., Shaltout, M., & Fekri, M. (2009). Astronomy, landscape and symbolism: A study on the orientations of ancient Egyptian temples. In J. A. Belmonte & M. Shaltout (Eds.), *In search of cosmic order, selected essays on Egyptian archaeoastronomy* (pp. 211–282). Cairo: Supreme Council of Antiquities.

Boutsikas, E. (2009). Placing Greek temples: An archaeoastronomical study of the orientation of ancient Greek religious structures. *Archaeoastronomy, 21*, 4–19.

Boutsikas, E., & Hannah, R. (2012). Aitia, astronomy and the timing of the Arrhephoria. *Annual of the British School at Athens, 107*, 233–245.

Boutsikas, E., & Ruggles, C. L. N. (2011). Temples, stars, and ritual landscapes: The potential for archaeoastronomy in ancient Greece. *American Journal of Archaeology, 115*(1), 55–68.

González-García, A. C., Belmonte, J., & Polcaro, A. (2016). A diachronic analysis of orientation of sacred precincts across Jordan. *Mediterranean Archaeology and Archaeometry, 16*(4), 133–141.

Hannah, R., & Magli, G. (2011). The role of the sun in the Pantheon design and meaning. *Numen, 58*(4), 486–513.

Kropp, A. J. M. (2009). The cults of Ituraean Heliopolis (Baalbek). *Journal of Roman Archaeology, 22*, 365–380.

Kropp, A. J. M. (2010). Jupiter, Venus and Mercury of Heliopolis (Baalbek). The images of the "triad" and its alleged syncretism. *Syria, 87*, 229–264.

Kropp, A. J. M., & Lohmann, D. (2011). 'Master, look at the size of those stones! Look at the size of those buildings!' Analogies in construction techniques between the temples at Heliopolis (Baalbek) and Jerusalem. *Levant, 43*, 38–50.

Lohmann, D. (2008). Recent architectural research on the planning and development of the sanctuary of Jupiter. In M. van Ess (Ed.), *Baalbek-Heliopolis. Results of archaeological and architectural research 2002–2005. Colloquium Berlin 2006* (pp. 153–160). Beirut: Direction general des Antiquites.

Lohmann, D. (2010). Giant strides towards monumentality: The architecture of the Jupiter Sanctuary in Baalbek/Heliopolis. *Bolletino di Archeologia, 1*, 29–30.

Magli, G. (2006). The acropolis of Alatri: Architecture and astronomy. *Nexus, 8*, 5–16.

Magli, G. (2015). *Archaeoastronomy. Introduction to the science of stars and stones*. New York: Springer.

Ritmeyer, L. (1992). Locating the original Temple mount. *Biblical Archaeology Review, 18*, 2445–2449.

Segal, A. (2013). *Temples and sanctuaries in the Roman East: Religious architecture in Syria, Iudaea/Palaestina and Provincia Arabia*. Oxford: Oxbow Books.

Chapter 10
Wind, Sand and Water: The Orientation of the Late Roman Forts in the Kharga Oasis (Egyptian Western Desert)

Corinna Rossi and Giulio Magli

Abstract The chain of late Roman fortified settlements built in the Kharga Oasis, in Egypt's Western Desert, represents an interesting case-study to analyse how the ancient Roman town planners interacted with the landscape. A peculiar feature of the site is the existence of a prevailing, north-westerly wind, and it is possible to identify the average azimuth of the wind by measuring the central axes of the half-moon shaped sand dunes which characterize the landscape. Using the methods of archaeo-astronomy, we compared these azimuths with the orthogonal layout of both the settlements and the agricultural installations and showed that these are oriented on the prevailing wind. A description and the possible implications of this 'weathervane orientation' are discussed in this paper.

Introduction

The northern outskirts of Kharga, one of the five major oases of Egypt's Western Desert, are punctuated by a scatter of Late Roman installations that survive in relatively good conditions, thanks to the dry desert environment and their remote position. This network of sites, possibly belonging to the re-organisation of the Empire's southern frontier triggered by Diocletian, was built along the most important caravan routes that met in the Kharga Oasis, a desert hub for trans-Saharan commercial movements. They appear to have played a combination of roles: besides controlling the routes, they exploited the agricultural and mining potential of the peripheral areas of the oases, and demonstrated the presence and power of Rome thanks to their imposing and aggressive appearance (Reddé 1991; Rossi 2013; Rossi and Ikram 2018: Chap. III.1; Tallet et al. 2012a).

C. Rossi (✉)
Department ABC, Politecnico di Milano, Milan, Italy
e-mail: corinna.rossi@polimi.it

G. Magli
Department of Mathematics, Politecnico di Milano, Milan, Italy
e-mail: giulio.magli@polimi.it

G. Magli et al. (eds.), *Archaeoastronomy in the Roman World*, Historical & Cultural Astronomy, https://doi.org/10.1007/978-3-319-97007-3_10

All these sites were endowed with large-scale agricultural systems, that were meant to make them self-sufficient (Bravard et al. 2016; Rossi 2016; Rossi and Ikram 2018: Chap. III.6; Tallet et al. 2011). The sites located on lower ground, closer to the centre of the oasis' depression, were served by wells, whereas the ones located on higher ground mainly relied on subterranean aqueducts of the type called *qanawat* or *manawir*. The subterranean aqueducts of Umm al-Dabadib were the first archaeological feature of the entire area to be described in some detail by the British geologist and explorer H. Ll. Beadnell, who worked with J. Ball at the Geological Survey of Egypt at the very beginning of the twentieth century (Ball 1900; Beadnell 1909).

In the majority of the Late Roman archaeological sites of north Kharga, the remains of the ancient fields still survive: the *centuriatio* can be clearly discerned from aerial and satellite images, and at ground level at dawn and sunset when the Sun is very low (Rossi 2016). The buildings belonging to this chain of installations share a significant number of architectural features, suggesting that the builders drew from a common set of models (Rossi and Ikram 2018: Chaps. III.1 and III.2). We will focus here on the problem of the orientation of these archaeological sites.

It is indeed well known that, according to a variety of ancient sources, Roman city and land planning involved procedures inherited from the Etruscans and closely connected with the celestial cycles. For instance, the town's foundation ritual is described by Roman historians (such as Varro, Plutarch and Pliny the Elder) as a rule directly inherited from the Etruscans' sacred books of the aruspexes, the *Disciplina*. A fundamental part of the aruspexes' duty was connected with the cosmic order, and, as a consequence, a role for astronomy is to be expected in the orientation of the Roman layouts. In the literature, a number of cases have been actually documented (Bertarione and Magli 2015; Espinosa Espinosa et al. 2016; González García and García Quintela 2014; González-García et al. 2014; Magli 2007, 2008; Rodrıguez-Anton et al. 2016) and, in particular, cardinal orientation was one of the possible choices.

In spite of this, attempts at establishing a common rule must comply with the practical mentality of the Romans, which in many cases overruled symbolic principles in favour of local (e.g. geomorphological) considerations. For instance, *centuriations* were clearly inspired by astronomical considerations (for instance that of Roman Carthage), but others were clearly inspired exclusively by the morphology of the terrain. One example is Luni, in Italy, in the Massa Carrara province. The land division there was to be planned in a territory that is confined between the (roughly parallel) natural borders of the Ligurian Sea and the Apuan mountain range. As a consequence, a grid that was not parallel to the shoreline would rapidly lead to many incomplete lots at both ends. The Roman planners thus opted for a purely practical solution and traced the main axis of the *centuriatio* along the natural constraint line. In the present paper, we shall show that the planning of the Roman settlements in Kharga also was very likely an example of orientation purely based on a practical solution: in this case, the need to take into account the direction of the prevailing wind.

The Late Roman Settlements in Northern Kharga

In comparison with the Nile Valley, the archaeological remains of the oases received little attention until the Egyptian Egyptologist Ahmed Fakhry embarked on a tour of the area in the 1940s. His reports were published in subsequent decades (Fakhry 1951, 1973), but his notes on the Kharga Oasis remain unpublished. A more systematic exploration began later, together with the first asphalted road, which reached Kharga in the 1960s. The archaeological sites located closer to the inhabited parts of the oasis were studied first (Cruz-Uribe 1988; Fakhry 1951; Reddé et al. 2004). They were followed by the areas located on the outskirts (e.g. Gascou et al. 1979; Ibrahim et al. 2008), more isolated sites (Rossi 2000; Rossi and Ikram 2010) and eventually by the most elusive remains, such as agricultural installations associated with these sites (Newton et al. 2013; Rossi 2016; Wuttmann et al. 2000). These studies slowly built up a coherent picture of the situation around the Kharga Oasis in the Roman period (Reddé 1991; Rossi and Ikram 2018: Chap. III.9; Tallet et al. 2012b; Wagner 1987).

Further work will be necessary in the future to address the earlier historical periods, which are still poorly represented. The history of the Kharga Oasis during the pharaonic period is, in fact, still obscure. Prehistory left abundant traces on the now-desert outskirts of the depression, which was once occupied by a large lake (Caton-Thompson 1952). The pharaonic remains, instead, are likely to have been concentrated in the lowest area of the depression and to have been buried by the continuous occupation of the greenest portion of the oasis, throughout the subsequent millennia. Only a few remains dating to this long period so far have been retrieved from the core of the oasis (Cappers et al. 2014), but the caravan routes that crossed the region offer substantial evidence on the existence of a network of transport and communications, suggesting a continuous and significant occupation of the area throughout all of that historical period (Ikram 2006; Ikram and Rossi 2004a; Rossi and Ikram 2002, 2013, 2015).

Late Period remains are limited, but significant: the earliest core of the Hibis Temple dates to this period, and indicates a strong interest in this area (Cruz-Uribe 1988; Winlock 1941). The scale of the Ptolemaic exploitation of the Western Desert oases is unclear, but might have been more substantial than previously thought (Gill 2016).

The oasis is clearly punctuated by well-preserved archaeological remains indicating that the Romans poured significant resources into the construction of settlements; in particular, a chain of military-looking installations was built in the Late Roman period. In northern Kharga they are, from north to south: the little forts of Qasr al-Gib and Qasr al-Sumayra (Ikram and Rossi 2004b), the fort of Mohammed Tuleib, the legionary fortress of al-Deir (Brones and Duvette 2008), and two similar, rather substantial installations: the Fort and the surrounding Gridded Settlement at Ain al-Lebekha (Rossi and Ikram 2010), and the Fort and the Fortified Settlement at Umm al-Dabadib (Rossi 2000; Rossi and Ikram 2006).

Near the central part of the oasis' depression, two more sites might belong to this chain: Qasr al-Nessima, and Qasr al-Baramoudy. They have never been studied and lie in an area that is not currently covered by good resolution satellite images. In general, they strongly resemble Umm al-Dabadib and Qasr al-Lebekha in terms of general layout, as they both consist of a sturdy central building surrounded by a dense settlement laid out in a grid pattern.

To the Roman period, broadly speaking, also belong three sites consisting of mudbrick enclosures surrounding sizeable stone temples, Nadura, Qasr al-Ghweita and Qasr al-Zayyan, that have been partly investigated (Cappers et al. 2014; Klotz 2009, 2013). In the south of the oasis, the most important Roman installation is the large site of Dush. This vast and stratified site had earlier origins: the nearby site of Ain al-Manawir appears to have been occupied from the Persian period onwards, and the fourth century AD Roman military base of Dush is likely to have re-used the structure of older fortified magazines (Reddé et al. 2004).

We shall now focus on the orientation of the Late Roman settlements belonging to the chain that was built in the northern and central portion of Kharga.

Orientation(s)

In Kharga, only small portions of the extant archaeological complexes have been excavated and surveyed in a detailed way. This is due to a combination of reasons: first of all, because large-scale excavation can be extremely complicated in such a sandy environment (e.g. Reddé et al. 1990: 284); then because the complex logistical conditions dictate the adoption of some methods instead of others (e.g. Fassi et al. 2015); and finally because of the lack of fixed points to geo-reference the surveys at the most remote sites. This means that we do not possess a large number of precise surveys, and that the orientations of several settlements and installations have so far been measured in an approximate way. The Hibis Temple represents an exception, as it has been the object of a careful conservation project.

We shall now focus on the chain of Late Roman fortified settlements that punctuate the Kharga Oasis, namely Qasr al-Gib, Qasr al-Sumayra, Muhammad Tuleib, Qasr al-Lebekha, Umm al-Dabadib, al-Deir, Qasr al-Nessima and Qasr al-Baramoudy. The general lack of precise measurements can be circumvented by using Google Earth. Accuracy of high-resolution Google Earth images is usually very good (Potere 2008; Pulighe et al. 2016); however, in this specific region the images, encompassing vast desert areas, are less defined, and the outlines of the few buildings that punctuate the landscape are slightly blurred. We therefore estimated the tolerance of our measurements as $\pm 1^{\circ}$, which is adequate for the current discussion (Fig. 10.1). All the structures are square or rectangular in plan, and therefore it is sufficient to measure the azimuth of one of their sides. The results are reported in Table 10.1.

The data cluster around true north, but we can exclude cardinal orientation as in this case the errors committed by the builders would have been too high. A key to the

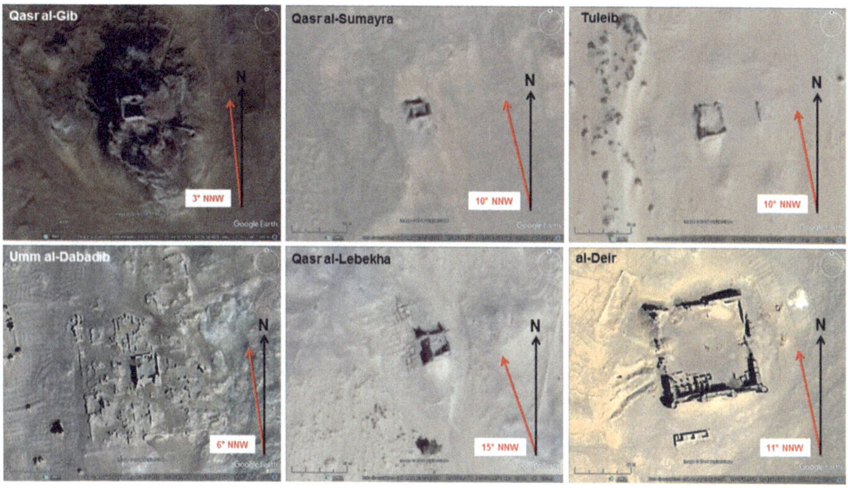

Fig. 10.1 Google Earth images of six Late Roman settlements in northern Kharga and their orientation

Table 10.1 Orientation of the Late Roman fortified sites of Kharga

Name	Azimuth of built-up areas	Approximate azimuth of barchan dunes
Qasr al-Gib	3°W	~10°W
Qasr al-Sumayra	10°W	~10°W
Tuleib	10°W	~10°W
Umm al-Dabadib	6°W	~7°W
Qasr al-Lebekha	15°W	~14°W
Al-Deir	11°W	~11°W
Qasr al-Baramoudy	10°E	~10°E
Qasr al-Nessima	7°E	~7°E
Dush	15°W	8–15 W
Dush Settlement	8°W	8–15 W

interpretation of these data is offered by the Fortified Settlement of Umm al-Dabadib and its associated cultivations (Fig. 10.2): the satellite images clearly show that the orientation of the Fortified Settlement and of the *centuriatio* is the same as the average orientation of the sand dunes. Dunes are shaped by the wind, and *barchan* (or half-moon shaped) dunes, as the ones at Dabadib, develop when the wind always blows from the same direction (Bagnold 1941: especially Chap. 14); it therefore possible to establish the average orientation of these natural features by taking the (of course approximate) azimuth of the central axis of the arch-shaped form. These dunes are persistent features of the landscape: the size of the area covered by them and the relatively slow speed at which they move indicate that the wind has been blowing in the same way for the last six thousand years (for a summary, see Rossi and Ikram 2018: 6–8).

Fig. 10.2 Google Earth image of the Fortified Settlement of Umm al-Dabadib and surroundings, showing the orientation of the built-up area, the *centuriatio*, the aqueducts and the sand dunes

The Fortified Settlement and its associated cultivations were therefore oriented *secundum naturam*: the natural feature to be taken into account here being the prevailing wind.

The entire Western Desert is actually shaped by the constant north-north-westerly wind: seen from the satellite, the entire region is crossed by an endless sequence of 'slanting', parallel lines of sand and dunes. The shape of the dunes reveals the wind pattern that, however, can be locally affected by the presence of changes of altimetry. By looking at the way sand accumulates, one can gain a fairly good idea of the local winds, and thus compare them with the orientation of the local settlements almost at each site (cf. Table 10.1).

The comparison with the forts' orientations, listed from north to south, clearly shows a close correlation, with one exception: Qasr al-Gib, the northernmost fort. This building is perched on a rock outcrop, in the midst of a rather rocky area; the orientation of this building is 3°W, and cannot be precisely matched with any topographic feature. The fort has a smaller satellite some 2 km to the south, Qasr al-Sumayra, which is also located far from any substantial accumulation of sand. The orientation of the latter, however, differs from the one of Qasr al-Gib and is, instead, the same of the large dunes in the plain which is about 10°W.

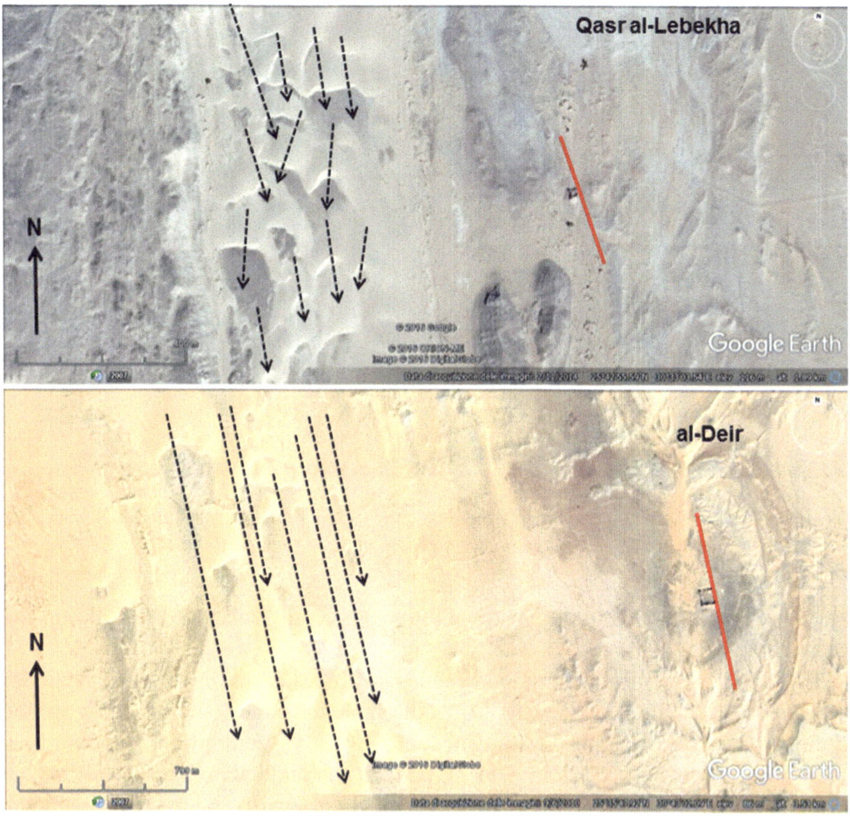

Fig. 10.3 Google Earth images showing the orientation of the two Late Roman settlements of Qasr al-Lebekha and al-Deir and the sand dunes that punctuate their surroundings

The Fort and the Gridded Settlement of Ain al-Lebekha follow a line pointing ca 15°W; the local chain of dunes roughly follows the same direction, although the varied shape of the dunes indicate the presence of turbulences and changes of direction of the prevailing wind, due to the proximity of the escarpment and to the presence of a high massif immediately to the west of the site (Fig. 10.3 top). The large legionary fortress of al-Deir is clearly aligned (ca 11°W) with the dunes that proceed undisturbed far to the west in the barren plain (Fig. 10.3 bottom).

A similar situation occurs for the fort at Muhammed Tuleib, some 8.5 km to the east. No dunes currently pass by Muhammed Tuleib, but the stripes and marks left on the surrounding landscape by the constant wind matches the orientation of the building at 10°W. This building contains the remains of an older mudbrick temple, of which, however, very little is known; the depleted remains of its eastern façade were incorporated in the outer wall of the new building, but the poor conditions of the remains do not allow a precise investigation of the relative orientation of the

Fig. 10.4 Google Earth images showing the orientation of the settlements of Qasr al-Nessima and Qasr al-Baramoudy, in central Kharga, in comparison with the local sand dunes

temple and the external building without proper archeological excavation (Ikram and Rossi 2007: 175–176).

The situation of Qasr al-Baramoudy and Qasr al-Nessima is interesting: the first is oriented 10°E and the second 7°E. They both lie at the edges of the deepest portion of the oasis' depression, near or surrounded by thick vegetation. The pattern of the sand shaped by the wind is less evident, and yet the undulating thick layer of sand that engulfs Qasr al-Baramoudy matches the slight eastward orientation of the main buildings. No sand dunes are close to Qasr al-Nessima but its orientation is very similar to that of Qasr al-Baramoudy (Fig. 10.4). Satellite images clearly shows that the orientation of the Baramoudy settlement appears to be slightly irregular and displaced a few degrees west, so that it does not match the orientation of the central building. If compared with the perfect alignment of the Fort with the Fortified Settlement at Umm al-Dabadib, this situation might indicate that the central building and settlement were not built together.

In the south of the oasis, the large site of Dush represents an interesting case. Differently from the Late Roman settlements in the north, here it is clear that the built-up area is the result of a long evolution that spans several centuries, at least from the first to the fifth century AD. The main building, reused by the Roman army in the fourth century AD, follows an orientation of ca 15°W; it contains (and is aligned to) the remains of a substantial early Roman stone temple. The nearby remains of another mudbrick temple are oriented along a similar direction, whilst the surrounding, slightly irregularly laid-out settlement follows a direction of ca 8°W (Fig. 10.5). The sand dunes that move south in the plain nearby follow directions less defined than in other areas but range between 8° and 15°W ca. The stratified nature of the site and the extended chronology of its occupation makes the interpretation of these remains less simple in comparison with the northern sites which were built in

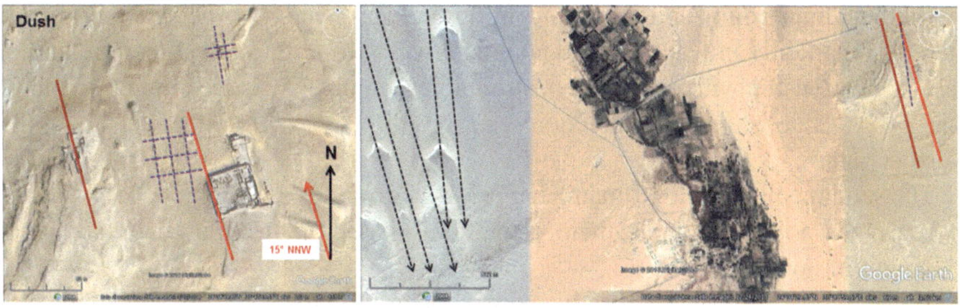

Fig. 10.5 Google Earth images showing the slightly different orientations of the various elements of the built-up area of Dush, in southern Kharga, in comparison with the local sand dunes

one phase and in the same historical period. At any rate, it is interesting to note that, in this case also, the orientations of the various components of this large and complex site match those of the local wind.

Finally, in central Kharga there are three sites that developed as mudbrick enclosures around pre-existing stone temples: Nadura, Qasr al-Ghweita and Qasr al-Zayyan. They are not included in Table 10.1 since it is clear that orientation of the temples—which might have depended on different factors—influenced that of the enclosures. For completeness, it is worth mentioning that their azimuths are easily measurable and essentially coincide with the one given by Belmonte and Shaltout (2006) for the temples, that is ~93°E for Qasr al-Ghweita, ~108°E for Nadura and 179°E for Qasr al-Zayyan.

Discussion: A 'Weathervane Orientation'?

The results presented above clearly point to the conclusion that the Late Roman settlements were planned taking into account the prevailing NNW wind.

The Roman architects were well aware of the importance of the local winds when they established the orientations of their towns: Vitruvius, for instance, dedicated an entire chapter of his Book I to the winds and their dangers, stressing the necessity to avoid that they would blow across the streets (*De Architectura* I, 6). He also provides a (rather unpractical) method to establish the best orientation based on the use of a gnomon to establish the north-south direction, followed by the division of the surrounding circle into 16ths, with the final aim to identify the eight regions of the main winds. The final recommendation is that "... the alignment of the streets and side streets ought to follow the angles between the regions of two different winds." (*De Architectura* I, 6.7, Rowland and Noble Howe 1999: 30).

In his Book XVIII, 'The Natural History of the Grain', Pliny the Elder also suggested an astronomical method for establishing the four cardinal points and, as a consequence, the direction from which the winds blow locally (*Naturalis Historia* XVIII, 76. (33.): The Theory of the Winds). All these complex passages to establish first the cardinal points and then to identify the local winds can be skipped in case there is a markedly prevailing wind coming always from the same direction, and in case its relationship with the broader astronomical context is deemed unimportant. In this case, establishing its direction can be directly achieved with a simple weathervane.

The settlements of Late Roman Kharga appear to follow a desert-adapted version of Vitruvius' principle, that is, their orientation is based on the wind, but their aim is to exploit it, not to avoid it.

Vitruvius underlined the disadvantages of the winds ("... cold winds are disagreeable, hot winds enervating, moist winds unhealthy ...", Book I, Chap. 6.1) and recommended "... shutting out the winds from our dwellings ..." (6.3; cf Rowland and Noble Howe 1999). This could be achieved by not laying down the streets facing the direction of the winds, and pointing the corners of the buildings in that direction, so that their force would be broken and dispersed (6.8). In Egypt, however, the northerly wind is generally welcome, as it provides relief from the summer heat; and even if in the desert it could be quite strong, it was perceived as a natural element to be exploited rather than avoided. Traditionally, the main doors of ancient Egyptian houses were located along the northern side, and this tradition appears to have continued well into the Roman times, as the houses in the nearby Dakhla Oasis attest (Boozer 2016: 158).

The Late Roman forts and settlements under study were built not to divert the wind, but to 'filter' it. Differently from the typical Egyptian houses, the main gate generally faced south, but this depended on a combination of reasons. Qasr al-Gib, for instance, was endowed with a single opening, the main gate facing south, which was disguised under a porch. It was a small solid building meant to act as a checkpoint along the caravan route heading north, and was not endowed with any visible opening probably to disguise from the distance its real size (Ikram and Rossi 2004b: 77). The Fort at Umm al-Dabadib, instead, was probably designed to convey an aggressive appearance to travelers coming from south, and the builders felt evidently free to take full advantage of its orientation: whilst the southern front was endowed with two towers, one gate and a small number of windows, the flat northern front was pierced by at least eighteen windows, one per room. They were all small, in order to reduce the heat and the sand that would otherwise sneak inside, but must have provided a welcome draught of cool air inside the building.

Unfortunately no other northern walls survive in good conditions in the other Late Roman forts, and it is difficult to tell whether the same system was adopted elsewhere. In fact, it should be noted that the choice to orient the walls of the settlement facing the northern wind took its toll on the constructions: after sixteen centuries, nearly all the east-west walls have been eroded away by the wind-blown sand. The northern faces of forts and settlements, if they still exist, are reduced to a battered and irregular mass of mudbricks, with two exceptions: Qasr al-Gib, located on a rock outcrop and evidently less exposed to the force of the heavier sand grains blowing at ground level, and the Fort Umm al-Dabadib, where the mass of the Fortified Settlement absorbed most of the damage and protected the central Fort.

Following Vitruvius' recommendation to align the corners of the buildings towards the main wind might have protected the wall facing north from the erosion, but would have fatally diverted the evening breeze. Clearly the long-term preservation of these constructions belongs to our realm of interest, whereas the ancient builders were rightly concerned with the immediate comfort of the inhabitants of their constructions.

Whether or not this 'weathervane orientation' was introduced by the Romans is impossible to tell, as the pre-Roman remains (besides the temples) in Kharga are too scant and poorly documented to draw any conclusion. In any case, the same alignment appears in a clear way in the layout of the agricultural system. The seven subterranean aqueducts that provided water to the cultivations at Umm al-Dabadib were quarried along the sides of the *wadis*, to avoid both the loose filling of the lowest part and the taller rocks on the upper part. Once the water reached the surface, it was channeled towards the fields by long open-air canals, constructed by taking advantage of the local topography, on sequences of mounds modeled by the erosion (thus also shaped by the constant wind). For this reason, the aqueducts also generally appear to follow a similar alignment if compared with the Fortified Settlement and the sand dunes. The orientation of the cultivations, clearly visible from the satellite images, perfectly matches that of the Fortified Settlement, and therefore of the dunes. This layout was clearly intentional: by following the topography of the *wadis* shaped by the erosion and of the plains bordered by the fields of dunes, the *centuriatio* filled the available space in the most effective way, by aligning the longest lines to the longest dimension of the terrain to be cultivated (cf. Fig. 10.2).

Concluding Remarks

In conclusion, the Late Roman settlements of north Kharga were designed to exploit in the best possible way the local environmental conditions in terms of air circulation and topography. The apparent absence of astronomical references in the orientation of these settlements and, on the opposite, the strong link with the local environment suggest that this Roman large-scale investment focused on the local conditions: the optimal exploitation of the local terrain appears to have been the main concern and the focus of the builders' efforts. This represents an indirect clue to the scope of this chain of settlements, evidently meant to firmly establish Roman control over this specific area.

Clearly, from the point of view of archaeoastronomy, the conclusions exposed here correspond to negative results, since we have shown that astronomical orientation was *not* used. However, negative results are of course a key ingredient of the scientific methodology of any discipline, and to reach our conclusions we fully applied the approach of modern archaeoastronomy, which takes into account all the aspects (natural, and man-made) of the landscape, not only of the sky.

References

Bagnold, R. A. (1941). *The physics of blown sand and desert dunes*. London: Methuen.

Ball, J. (1900). *Kharga Oasis: Its topography and geology*. Cairo: National Printing Department.

Beadnell, H. L. (1909). *An Egyptian Oasis; An account of the Oasis of Kharga in the Libyan desert, with special reference to its history, physical geography, and water-supply*. London: J. Murray.

Belmonte, J. A., & Shaltout, M. (2006). On the orientation of ancient Egyptian temples: (2) New experiments at the oases of the Western Desert. *Journal for the History of Astronomy, 37*(2), 173–192.

Bertarione, S., & Magli, G. (2015). Augustus' power from the stars and the foundation of Augusta Praetoria Salassorum. *Cambridge Archaeological Journal, 25*, 1–15.

Boozer, A. L. (2016). *Amheida II. A Late Romano-Egyptian House in Dakhla Oasis: Amheida House B2*. New York: New York University Press.

Bravard, J.-P., Mostafa, A., Garcier, R., Ballet, P., & Tallet, G. (2016). Rise and fall of an Egyptian Oasis: Artesian flow, irrigation soils and historical agricultural development in El-Deir, Kharga depression, Western Desert of Egypt. *Geoarchaeology, 31*(6), 467–486. https://doi.org/10.1002/gea.21566.

Brones, S., & Duvette, C. (2008). Le fort d'El-Deir, oasis de Kharga: 'État des lieux' architectural et archéologique'. *Bulletin de l'Institut Français d'Archéologie Orientale, 107*, 5–41.

Cappers, R., Heinrich, F., Kaaijk, S., Fantone, F., Darnell, J., & Manassa, C. (2014). Barley revisited: Production of barley bread in Umm Mawagir. In K. Accetta, R. Fellinger, P. Lourenço Gonçalves, S. Musselwhite, & W. P. van Pelt (Eds.), *Current research in Egyptology 2013: Proceedings of the fourteenth annual symposium* (pp. 49–63). Oxford: Oxbow.

Caton-Thompson, G. (1952). *Kharga Oasis in Prehistory*. London: The Athlone Press.

Cruz-Uribe, E. (1988). *Hibis temple project*. San Antonio: Van Siclen Books.

Espinosa Espinosa, D., González García, A. C., & García Quintela, M. V. (2016). On the orientation of two Roman towns in the Rhine area. *Mediterranean Archaeology and Archaeometry, 16*(4), 233–240.

Fakhry, A. (1951). *The Necropolis of Bagawat in Kharga Oasis*. Cairo: Government Press.

Fakhry, A. (1973). *The Oases of Egypt 1: Siwa Oasis and II: Bahariyah and Farafra Oases*. Cairo: American University in Cairo Press.

Fassi, F., Rossi, C., & Mandelli, A. (2015). Emergency survey of endangered or logistically complex archaeological sites. *International Archives of the Photogrammetry, Remote Sensing and Spatial Information Sciences, XL-5 and W4*, 85–91. https://doi.org/10.5194/isprsarchives-XL-5-W4-85-2015.

Gascou, J., Wagner, G., & Grossmann, P. J. (1979). Deux voyages archéologiques dans l'oasis de Khargeh. *Bulletin de l'Institut Français d'Archéologie Orientale, 79*, 1–20.

Gill, J. C. R. (2016). *Dakhleh Oasis and the Western Desert of Egypt under the Ptolemies*. Oxford: Oxbow (Dakhleh Oasis Project Monograph 17).

González García, A. C., & García Quintela, M. V. (2014). The 1st of August at Lugdunum: Astronomy and imperial cult in Galia. *Mediterranean Archaeology and Archaeometry, 14*(3), 83–91.

González-García, A. C., Rodríguez-Antón, A., & Belmonte, J. A. (2014). The orientation of Roman towns in Hispania: Preliminary results. *Mediterranean Archaeology and Archaeometry, 14*(3), 107–119.

Ibrahim, B. A., Dunand, F., Heim, J.-L., Lichtenberg, R., & Hussein, M. (2008). *Materiel Archeologique et les Restes Humains de la Necropole d'Ain el-Labakha*. Paris: Cybele.

Ikram, S. (2006). Paths, petroglyphs, and piety: The North Kharga Oasis survey 2006 season. *Bulletin of the American Research Center in Egypt, 190*, 18–23.

Ikram, S., & Rossi, C. (2004a). A new early dynastic *Serekh* from the Kharga Oasis. *Journal of Egyptian Archaeology, 90*, 211–215.

Ikram, S., & Rossi, C. (2004b). North Kharga Oasis survey 2001–2002. Preliminary report: Ain Gib and Qasr el-Sumayra. *Mitteilungen des Deutschen Archäologischen Instituts Kairo, 60*, 69–92.

Ikram, S., & Rossi, C. (2007). North Kharga Oasis survey 2004 preliminary report: Ain el-Tarakwa, Ain el-Dabashiya and Darb Ain Amur. *Mitteilungen des Deutschen Archäologischen Instituts Kairo, 63*, 167–184.

Klotz, D. (2009). The cult-topographical text of Qasr el-Zayyan. *Revue d'Égyptologie, 60*, 17–39.

Klotz, D. (2013). Yale University Nadura Temple Project: 2009 Season. In R. Bagnall, P. Davoli, & C. Hope (Eds.). *The Oasis papers 6. Proceedings of the sixth conference of the Dakhla Oasis project* (pp. 297–304). Oxford: Oxbow (Dakhla Oasis Project Monograph 15).

Magli, G. (2007). Non-orthogonal features in the planning of four ancient towns of Central Italy. *Nexus, 9*, 71–92.

Magli, G. (2008). On the orientation of Roman towns in Italy. *Oxford Journal of Archaeology, 27* (1), 63–71.

Newton, C., Whitbread, T., Agut-Labordère, D., & Wuttmann, M. (2013). L'agriculture oasienne à l'époque perse dans le sud de l'oasis de Kharga (Égypte, Ve-IVe s. AEC). *Revue d'Ethnoécologie, 4*, 2–20.

Pliny the Elder, *Naturalis Historia* (first century AD).

Potere, D. (2008). Horizontal positional accuracy of Google Earth's high-resolution imagery archive. *Sensors, 8*, 7973–7981.

Pulighe, G., Baiocchi, V., & Lupia, F. (2016). Horizontal accuracy assessment of very high resolution Google Earth images in the city of Rome, Italy. *International Journal of Digital Earth, 9*, 342–362.

Reddé, M. (1991). A l'ouest du Nil: Une frontière sans soldats, des soldats sans frontière. In V. A. Maxfield & M. J. Dobson (Eds.), *Roman frontier studies 1989. Proceedings of the XV international congress of Roman frontier studies* (pp. 485–493). Exeter: University of Exeter Press.

Reddé, M., Dunand, F., Lichtenberg, R., Heim, J.-L., & Ballet, P. (1990). Quinze années de recherches françaises à Douch. Vers un premier bilan. *Bulletin de l'Institut Français d'Archéologie Orientale, 90*, 281–301.

Reddé, M., Ballet, P., Lemaire, A., & Bonnet, C. (2004). *Kysis. Fouilles de l'Ifao à Douch (1985–1990)*. Cairo: Institut Français d'Archéologie Orientale (Documents de Fouilles de l'IFAO 42).

Rodrıguez-Anton, A., Belmonte, J. A., & Gonzalez-Garcıa, A. C. (2016). Astronomy in towns? An archaeoastronomical approach to the Roman urbanism highlights on Spanish astrophysics IX. In S. Arribas, A. Alonso-Herrero, F. Figueras, C. Hernández-Monteagudo, A. Sánchez-Lavega, & S. Pérez-Hoyos (Eds.), *Proceedings of the XII scientific meeting of the Spanish Astronomical Society held on July 18–22, 2016, in Bilbao, Spain* (pp. 760–765).

Rossi, C. (2000). Umm el-Dabadib, Roman settlement in the Kharga Oasis: Description of the visible remains. With a note on 'Ayn Amur'. *Mitteilungen des Deutschen Archäologischen Instituts Kairo, 56*, 235–252.

Rossi, C. (2013). Controlling the borders of the empire: The distribution of Late-Roman 'forts' in the Kharga Oasis. In R. Bagnall, P. Davoli, & C. Hope (Eds.), *The Oasis papers 6, Proceedings of the sixth conference of the Dakhla Oasis project* (pp. 331–336). Oxford: Oxbow (Dakhla Oasis Project Monograph 15).

Rossi, C. (2016). Italian mission to Umm al-Dabadib, season 2014 – preliminary report. *Mitteilungen des Deutschen Archäologischen Instituts Kairo, 72*, 153–172.

Rossi, C., & Ikram, S. (2002). Petroglyphs and inscriptions along the Darb Ain Amur, Kharga Oasis. *Zeitschrift der Ägyptischen Sprache, 129*, 142–151.

Rossi, C., & Ikram, S. (2006). North Kharga Oasis Survey 2003 preliminary report: Umm el-Dabadib. *Mitteilungen des Deutschen Archäologischen Instituts Kairo, 62*, 279–306.

Rossi, C., & Ikram, S. (2010). North Kharga Oasis survey 2007 – preliminary report: Ain Lebekha and Ain Amur. *Mitteilungen des Deutschen Archäologischen Instituts Kairo, 66*, 235–242.

Rossi, C., & Ikram, S. (2013). Evidence of desert routes across northern Kharga (Egypt's Western Desert). In H. Riemer & F. Förster (Eds.), *Desert road archaeology in Western Sahara* (pp. 265–282). Köln: Heinrich-Barth-Institut.

Rossi, C., & Ikram, S. (2015). New Kingdom activities in the Kharga Oasis: The scribe Userhat travels westwards. *Journal of Egyptian Archaeology, 100*, 476–480.

Rossi, C., & Ikram, S. (2018). North Kharga Oasis survey: Explorations in Egypt's Western Desert. Leuven: Peeters (British Museum Publications on Egypt and Sudan 5).

Rowland, I. D., & Noble Howe, T. (Eds.). (1999). *Vitruvius, ten books on architecture*. Cambridge: Cambridge University Press.

Tallet, G., Bravard, J.-P., & Garcier, R. (2011). L'eau perdue d'une micro-oasis. Premiers résultats d'une prospection archéologique et géoarchéologique du système d'irrigation d'El-Deir, oasis de Kharga (Égypte). In C. Abadie-Reynal, S. Provost, & P. Vipard (Eds.), *Histoire des Réseaux d'Eau Courante dans l'Antiquité – Réparations, Modifications, Réutilisations, Abandon, Récupération* (pp. 173–188). Rennes: Presses Universitaires de Rennes (Collection Archéologie et Culture).

Tallet, G., Gradel, C., & Guédon, S. (2012a). 'Le site d'El-Deir, à la croisée des routes du désert occidental: Nouvelles perspectives sur l'implantation de l'armée Romaine dans le désert Égyptien. In P. Ballet (Ed.), *Grecs et Romains en Égypte. Territoires, Espaces de la Vie et de la Mort, Objets de Prestige et du Quotidien. Actes du Colloque International de la SFAC, Paris, 15 Novembre 2007* (pp. 75–92). Cairo: Institut Français d'Archéologie Orientale.

Tallet, G., Bravard, J.-P., Guédon, S., & Mostafa, A. (2012b). The survey project at El-Deir, Kharga Oasis: First results, new hypotheses. In R. S. Bagnall, P. Davoli, & C. A. Hope (Eds.), *Oasis papers 6. New perspectives on the Western Desert of Egypt* (pp. 349–361). Oxford: Oxbow Books.

Vitruvius, *De Architectura* (first century BC).

Wagner, G. (1987). *Les Oasis d'Égypte*. Bibliothèque d'Étude 100. Cairo: IFAO.

Winlock, H. E. (1941). *The temple of Hibis in el-Khargeh oasis. Volume 1: The excavations*. New York: Metropolitan Museum of Art.

Wuttmann, M., Gonon, T., & Thiers, M. (2000). The Qanats of 'Ayn-Manâwîr (Kharga Oasis, Egypt). *Journal of Achaemenid Studies and Researches, 1*(1).

Part V
Roman Virtual Archaeology and Archaeoastronomy

Chapter 11
The Mausoleum of Theodoric: Archaeoastronomy, Numbers, Geometry and Communication

Manuela Incerti, Gaia Lavoratti, and Stefania Iurilli

Abstract The following paper focuses on the Mausoleum of Theodoric (520 ca.), one of Ravenna's Byzantine monuments and a UNESCO heritage site, presenting the results of different phases of research that begun in 2015. Starting from the instrumental survey carried out with laser-scanner and digital photogrammetry technology, the unit of measurement and the geometric properties of the decagonal shape of the design of this singular two-level building were analysed. The archaeoastronomical study has highlighted possible meanings of the orientation of the building and the positioning and sizing of small wall openings. Finally, a 3D model was developed from the survey data to verify the astronomical phenomena and to aid in the multimedia communication of the scientific content. It is increasingly clear how virtual models, both interactive and non-interactive, constitute an important edutainment tool. This element is indispensable to the development of contemporary methods of dissemination for the fruition of cultural sites and artifacts.

Introduction: The Foundation and the Main Topics

The historian known as *Valesiano* documents that the Mausoleum of Theodoric was commissioned by Theodoric himself before his death on AD 30 August 526 (Muratori 1738). Theodoric (Teodorico) was born around 454. At the young age of 12 he was sent to Constantinople as a hostage, and remained at the court of Leo I the Thracian until 472. Scholars do not agree on the terms and type of education he received in the East; however, it is undeniable that during his kingdom he showed great attention to architecture. This is testified by the restoration of ancient buildings in Rome and the construction of new buildings in Verona, in Pavia and, above all, in Ravenna.

The Mausoleum was developed on two levels: the ground floor has a decagonal plan in external profile and a Greek cross interior, while the upper floor has a

M. Incerti (✉) · G. Lavoratti · S. Iurilli
Department of Architecture, University of Ferrara, Ferrara, Italy
e-mail: icm@unife.it; gaia.lavoratti@unife.it; stefania.iurilli@unife.it

G. Magli et al. (eds.), *Archaeoastronomy in the Roman World*, Historical & Cultural Astronomy, https://doi.org/10.1007/978-3-319-97007-3_11

decagonal exterior and a circular interior space. Like all monuments in Ravenna, the building has been the object of specialized studies and surveys (Bovini 1977; Gotsmich 1958; Guberti 1952; Haupt 1913; Heidenreich and Heinz 1971; Johnson 1988). In its long history, the small, central plan Mausoleum has been the object of multiple transformations and restorations, such as those of the eighteenth, nineteenth and twentieth centuries (Conti and Berti 1997; Guberti 1952: 8–19). The last interventions date back to 1977 (Bovini 1977: I–XV; Piazza 2013: 84–86; from the same volume see Novara 2013: 111–116) and 1998, the year in which the restoration of the stone of Aurisina took place (Bevilacqua et al. 2003; Piazza et al. 1998).

In the present study deeper discussion will relate to elements of the architecture and topics concerning the form and orientation of the building. It is thus particularly important to verify the authenticity and the dating of the elements involved in the analysis to avoid erroneous interpretation of the data.

A Description of the Mausoleum

The Question of Its 'Unfinished' Nature

Some small arches appear on the external face of the second floor, which may hint at the past presence of a loggia, perhaps lost or never finished. In this regard, the question of the 'unfinished' and the possible different dating of the two levels introduced by some authors does not appear to interfere with our observations. All reconstructions hypothesised for the second floor, amongst which one must remember the extremely accurate and sophisticated one by De Angelis d'Ossat (1962, with very accurate graphics), never involve the openings but only address the presence and shape of the portico, which is lower compared to the system of windows.

The Flooring

The current flooring of both rooms is certainly not original: in 1557 Leandro Alberti mentioned traces of a mosaic floor, evidently on the upper level of the building, as the bottom was buried underground (Fagiolo 1972: 148–149). Regarding the progress of the flooring element, historians have reported a major failure of the ground on the eastern side, which led to a drop of 14 cm in the ground floor and a 6 cm drop of the upper floor. The difference in height between the two levels has led scholars to believe that an initial failure occurred during the construction of the ground floor. For this reason, the upper floor was probably put in place 'levelling' the already installed plan, which however, later experienced another slight lowering.

The existing pavements were put in place during the works of the biennium, in 1975–1977 (Novara 2013: 116). The current floors of the two levels are more or less horizontal (with an incline of a few centimetres). One can still see the signs of the collapse by looking at the slight inclination of the band present in the tambour of the dome (the quotes and sources of the surveys can be found in Guberti (1952: 37, 56–58).

The Small Apse

On the eastern side of the top space there is a small apse, whose function many historians have questioned: its height cannot accommodate an altar or an officiant or even the great porphyry sarcophagus (today placed at the center of the space). On the keystone of the arch is a large Latin cross, the only sculptural element of the interior, highlighting its relevance in the project. The small space, whose floor was slightly lower than that of the rest of the room is, according to scholars, contemporary with the building (De Angelis d'Ossat 1962; Messina 1980: 128–129).

The Sarcophagus

According to tradition, the remains of King Theodoric were conserved in the great sarcophagus of red porphyry measuring 305 × 190 × 101 cm. The tub is characterized by four rings on the top edge and two lion's heads at the bottom center of the side faces. The sarcophagus' troubled history has been well documented, its movements traced by Ambrogi (1995: 109–111) recalling its relocation to the site in 1913.

There is no certainty regarding the original orientation of this object, which is, however, considered by scholars to be consistent with the building, and originally arranged in an east-west direction (the current one).

The Small Windows

The wall of the ground floor has a thickness of about 140 cm and is pierced by six splayed narrow slits arranged on three sides (two on the north wall, three on the east wall and two on the south wall) with approximately horizontal intrados. Their sizes vary in width from 11 to 25 cm, and in height between approximately 40 and 70 cm. The decagonal part of the upper level presents a central receding band, about 77 cm thick and perforated by 11 windows. Arranged approximately in the directions north-south, east-west with the two diagonals at 45° (directions of the compass rose), the small openings have dimensions that vary from 40 cm in height for the windows on the north-south axis, to 62 cm on the diagonal axes. These windows are almost unanimously considered contemporary with the founding of the building, excluding the rectangular south-western one which was clearly enlarged at a later date (Guberti 1952: 94; De Angelis d'Ossat 1962: 59).

Keeping the axes of the openings described above in mind, the internal lighting system of the two rooms, formed by 17 openings (11 + 6) can be traced back to 12 different directions: 5 in the lower deck and 8 in the upper one. Of these, only one—the eastern one—follows the trend of and lower system. Below the windows is a protruding band (of about 8 cm) on which inscriptions laid out on three different levels have recently been found (Novara 2013: 116, see also 85; Piazza 2009). These were investigated and restored in 2012 (but the results have yet to be published).

The Dome

The great monolith that covers the building has also been subject to a great number of specialized studies, which have investigated physical, technological, figurative, historical and design aspects (e.g. see Bianco Fiorin 1993; Dyggve 1957; Fagiolo 1972; Tabarroni 1973).

The inner diameter is about 925 cm and the height on the springing is about 190 cm. A large crack, which popular tradition blames on a bolt of lightning, marks the southern side where a lighthouse was built adhering to the building.

Twelve protruding elements with triangular perforations are present on the outer edge of the roofing, conveying the image of a 'royal crown'. Historians have often questioned the real function of these elements and their figurative origin (Fagiolo 1972). The assumption is that they were used for the passage of cables and ropes necessary for the positioning of the roof, as hypothesised by Antonio da Sangallo in a previously published drawing (see Heidenreich and Heinz 1971: 63, Fig. 65), may be considered unfounded because of the enormous weight of the monolith and the common technical operations of the time (Tabarroni 1973). What all scholars emphasize is the lack of regularity in the arrangement of the dodecagon traced by the protruding elements, for it is not aligned with any of the geometries of the building. The monolith is in fact slightly rotated in relation to the main axis of the building, which has led to the unanimous conclusion of a faulty, unfixable installation due to the creation of the dangerous lesions on the southern side.

The names of the apostles and evangelists are inscribed on the vertical faces of the elements in a sequence (from the door, clockwise): Lucas, Marcus, Mathias (?), Matteus, Felippus, Johannes, Jacobus, Andreas, Paulus, Petrus, Simeon, Thomas. The reasons for this particular sequence have been widely investigated (Fagiolo 1972; Heidenreich and Heinz 1971; Tabarroni 1973). All elements are finished with a gable roof, almost simulating a small sarcophagus, except for one: that of Petrus has a flat roof. This has led researchers to believe in the existence of a terminal element made of a different—and perhaps more valuable—material (which then went missing), highlighting the figure of Petrus, the founder of the Church (Tabarroni 1973: 141).

The Architectural Survey

The architectural survey was carried out by M. Incerti and P. Lusuard with a Faro focus3d scanner. Thirty different survey stations were established covering the interior and exterior of the building. Individual clouds were registered with the aid of spherical targets (software for data management *Scene 5.3*, data elaborated by M. Incerti).

At a later date, two different photographic campaigns were carried out for the reconstruction of the three-dimensional textured model (M. Incerti): the first relating to the exterior, the second to the interior. The exterior shots were taken with a compact Lumix DMC-TZ7. The interior shots, due to matters of critical illumination, were produced with a digital SLR camera on a tripod. The upstairs photography was particularly problematic, as the view was obstructed by a railing which inevitably projected into the wall surface. It was also difficult to address the problem of backlighting generated by the perforated doors with cross motifs, as was the issue of artificial lighting, which created disruptive shadows.

Survey Drawings: Methods and Procedure

The thirty clouds available produced a dense pointcloud with rather limited bands of occlusion (the absence of this data is only found in small portions of the building where the height of the scanner failed to balance the overhanging parts of the structures).

For the creation of the two-dimensional canonical drawings (plans, elevations and sections), used to effectively describe dimensions and geometries, the point cloud model was divided by horizontal and vertical planes. A thin slice (thickness of 1 cm) was extracted from the cloud for each cut-plane as well as high definition screenshots. By importing the slice with vectorial software by interpolation of the points on a 1:1 scale, and exported for the realization of definitive raster images 1:50 scale. The choice of using a 'slice' of such reduced points yet still obtain a sufficiently detailed section was possible thanks to the particular density of the pointcloud, which provided a high degree of detail even on particularly elaborate portions such as the shell decorations on the interior brackets. The screenshots, mosaicized in order to achieve a high definition end result, allowed accurate control over the size and even deformations of elevation orthophotos produced by the digital photomodelling software, enabling the correct adjustments of projection elements.

The final elaborates are therefore the result of the overlapping of parts in section and projection obtained through the procedures described above. This work format, now well established in the scientific world, ensures greater metric control of the architecture (through comparison and contamination of drawings obtained through different processes, distinct survey campaigns and different instruments). It also allows for detailed graphics containing material and chromatic information that a traditional survey would not have been able to capture.

Archaeoastronomical Analysis

The Orientation

The Mausoleum has also been subject of archaeoastronomical research conducted by Giuliano Romano, who measured its orientation (Azimuth 84.5°; Romano 1995). The building is rotated by 5.5° compared to the equinoctial direction, which should not be overlooked during the alignment operations. Despite this apparent irregularity and approximation of the directions of the axes, an in-depth study of the consequences of these data seemed of interest.

Following the correction of the slight rotation, the survey methodology involved specially processed survey drawings. By overlaying graphics to the four main astronomical directions (solstices and equinoxes), the small windows placed in the 45° directions were not found to be perfectly aligned in relation to the center. Despite this, it is clear that these windows allow the entry of light during the two solstitial dates.

The Windows

Only three of the seventeen windows, those on the north side (two downstairs and one on the top floor) do not receive significant sunlight. All of the others are involved in important moments of the astronomical year. The behaviour of sunlight on horizontal and vertical surfaces was analyzed through plans and sections. Height and azimuth angles were traced to the ephemeris through specific software. Among the phenomena we noted (Fig. 11.1) that:

1. The rising Sun entered the cross-shaped window (second floor) on the days of the equinoxes, illuminating the previously mentioned thin band with painted writing on the axis of the cell. On the day of the summer solstice, about an hour after sunrise, the spot of sunlight passed over the stone sarcophagus.
2. The Sun entered the four narrow windows on the door (second floor) at sunset on the days of the equinoxes, illuminating the same scripted band. For other examples, see Incerti et al. (2016).

Phenomena of this kind may have been used for ritual purposes (see Fagiolo 1972, and the chapter by Hannah in this book), but also to mark the advent of a particular date in the year, or for the computation of time: in other words, to indicate a precise moment of the day (sunset, in this case).

Fig. 11.1 Photographs of the effects of light (21/3 sunset, 21/6 sunrise)

The Dome

The results of our survey allowed us to verify that the arrangement of the protruding elements does not follow the directions of the decagonal geometry, but the cardinal directions with rather accurate approximation. The item marked with the name of Petrus (the only one with a flat roof), is aligned south. Aligned to the east is Jacobus, to the west is Lucas, and to the north is Matteus. This azimuth value is a possible explanation for what scholars consider an 'executional mistake' since the upper elements appear inconsistent with the main axes of the plant.

The Interactive Models for the Dissemination of the Research Project

The above study of Theodoric's Mausoleum and the instrumental survey that supports it, have translated into a multitude of results and materials of a different nature. From the massive amount of data collected, new information has emerged regarding the geometrical and archaeological characteristics of the building. The issue of disseminating and communicating the results of the research is a theme that our group has faced for some years. We have concentrated on the production of explorable and electronically searchable digital models as complementary and heterogeneous containers of information.

The 3D digital model made for the Ravenna Mausoleum can be explored in dynamic perspective on screen. This constitutes a visual support that provides the user with multiple information about the object's morphology: its colours, the materials, its state of conservation and much more. It can also be used as a visual database, useful in systematizing and making use of data beyond the range of the naked eye (dimensional data, geometrical relationships between elements, archaeoastronomical analyses, wall stratigraphy, external metadata such as video and Multimedia, etc. . . .). The interrogation of the model and reasoned structuring of information according to different levels of depth, facilitates the understanding of complex phenomena for the recipient of the information.

Starting from the pointcloud from the digital survey, a 3d model of the entire building was created, both external and internal, in order to allow a direct visualization of the light phenomena affecting the spaces on particular dates of the year. The model, designed to be optimized for real-time applications, is a textured quadrangular mesh (*quad-modelling*), texturized with *UV mapping* starting from the orthophotos extracted from the SFM survey. This model, oriented and placed in a Cartesian space for reference, has been linked to a directional light simulating the parallel rays of a source similar to the Sun, and is therefore best suited to reproduce the Sun's movement within the Mausoleum. The light has been assigned an animated path that reproduces the Sun's movement on the ecliptic, where each key movement on the animation (*keyframe*) was created by parameterizing the values derived from the calculation of

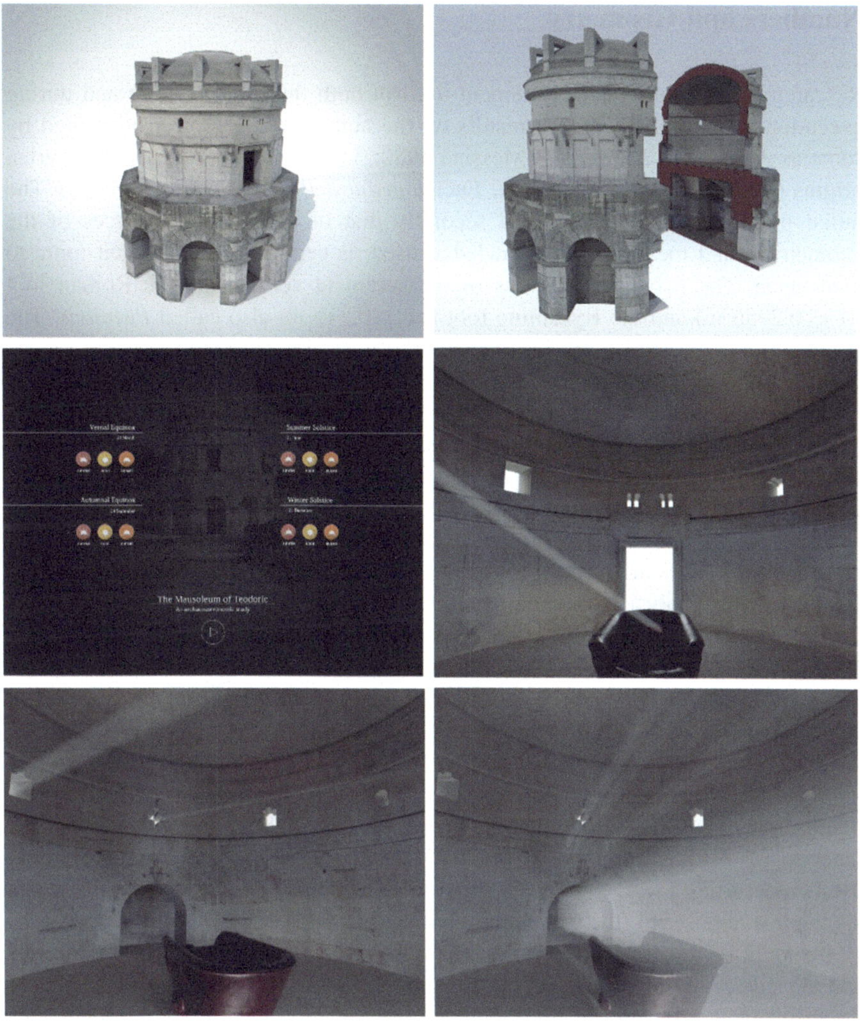

Fig. 11.2 Model and rendering of the building

the ephemeris at significant times. In particular, the exact time of sunrise has been entered as the starting point, and sunset as the end of the path. This time span was further subdivided into half hour intervals. Intermediate times result automatically from the data provided: the construction phase of the model thus becomes a test and comparison of the calculations previously made. The procedure was repeated for four remarkable dates (solstices and equinoxes). This model has been used as a kind of virtual laboratory for observing the effects of light within the burial cells in an ideal condition, since the light has no obstacles and external elements which, at present, obscure the sunrays (Fig. 11.2).

Numbers and Geometry

Research on the units of measurement used in both the project phase and during execution, can yield interesting results on the author of the project, identified by some as Aloisio—o Aloiosus—(Messina 1980: 33), an architect of debated Syrian origins (V. Aloisio and A. Iacobini, *Enciclopedia dell'Arte Medievale*, 1991). The initial problem of authorship, and secondly that of the possible sources of the geometrical and measurement knowledge used, is certainly an important topic to investigate. The two possible units of measurement verified are the Roman foot (rf = 0.2956 m) and the Byzantine foot (bf = 0.315 m, also called *Parmac*). The theme of the measurement of the Byzantine foot has been tackled in various papers (Ousterhout 2008: 75–76; Schilbach 1970, 1991; Underwood 1948) from which we extrapolate the values 0.312 m and 0.315 m (Martini 1883: 178). Throughout the research, both of these measurements were tested, with the result that the second value gave more 'whole' figures. The question of measurements, however, cannot be treated separately from the geometrical knowledge of the time.

The graphics elaborated by the instrumental survey made it possible to detect the presence of a geometrical design that led the metric control of the investigation. Beginning our analysis from the ground floor, the plan is based on a series of circumferences with a 'whole' radius measurement in which concentric decagons are inscribed (Figs. 11.3 and 11.4). The diameter of the circle in which the decagon is inscribed measures (Figs. 11.3 and 11.4) 45 Byzantine feet (bf), but also 47.92 Roman feet (rf), so almost 48 rf, two interesting measurements from a metrological analysis. Continuing with the measurements of the other decagons, we find that the internal line of the niches on the external side corresponds to the decagon inscribed in the circle with a diameter of 35 bf, and the diameter of the inscribable circle in the inner space of the ground floor (which can be traced back to the decagonal figure itself) measuring 25 bf. Finally, the thickness of the walls in the direction of the apothem is almost 9bf (the exact measurement is 8.9 bf).

It should be remembered that the relationship that binds the radius of the circle and the side of the inscribed regular decagon within is the irrational number 0.618, the result of the division of a unitary segment 'in extreme and mean ratio'. This numerical relationship between the parts of a segment, already present in *Elements* by Euclid (Book VI, Theorem VI, 30; Herz-Fischler 1998: 14), makes it clear that if the side of a decagon has a whole measurement (integer), the radius of the circumscribed circumference cannot have the same characteristic, and vice versa. With this binding condition comes the difficulty of calculating its area. Given the presence of an irrational number, the measurement of its surface has been subject to approximations such as those developed by Heron (Metric I, 23¸ Herz-Fischler 1998), whose formula $L^2 \times 15/2$ tries to be as rigorous as possible: $L^2 \times$ *fixed number of decagon* (the relationship between the apothem and the side), $L^2 \times 7.694$. Another numerical relationship used for the fixed number of the decagon is 38/5 (7,6) which comes closer to the exact value of 7.694 (Herz-Fischler 1998: 110).

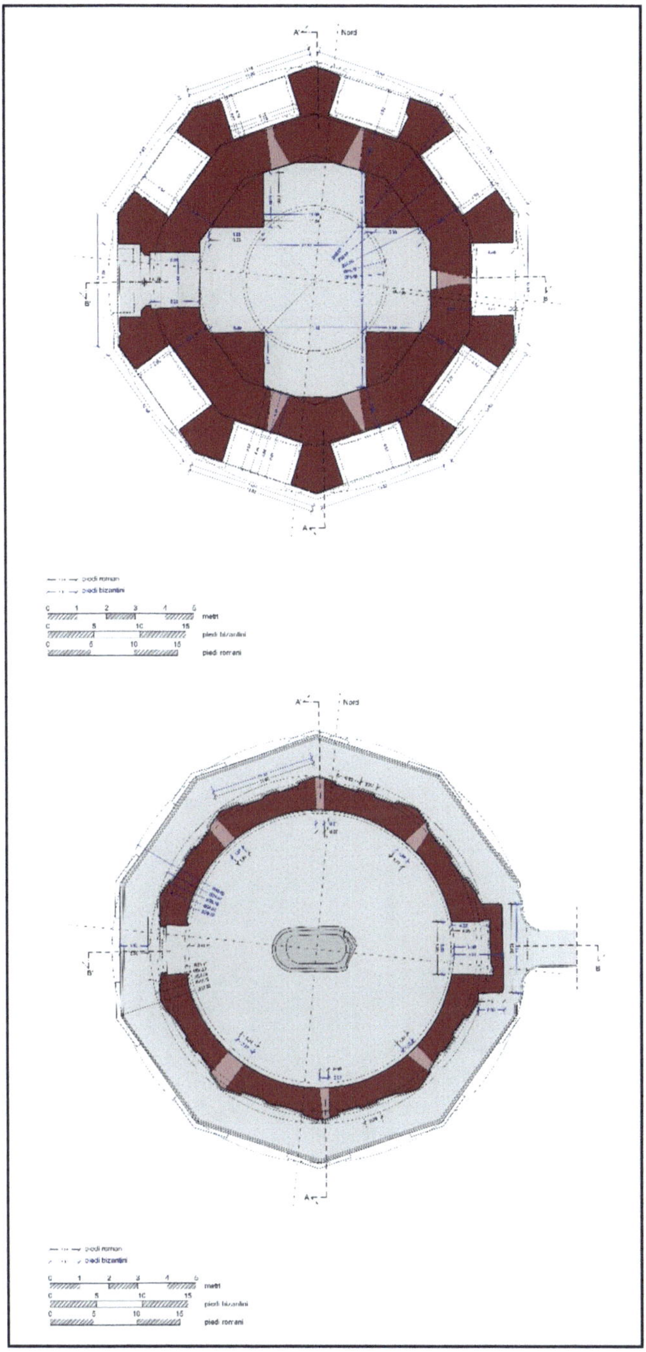

Fig. 11.3 Plan of the first and second floor

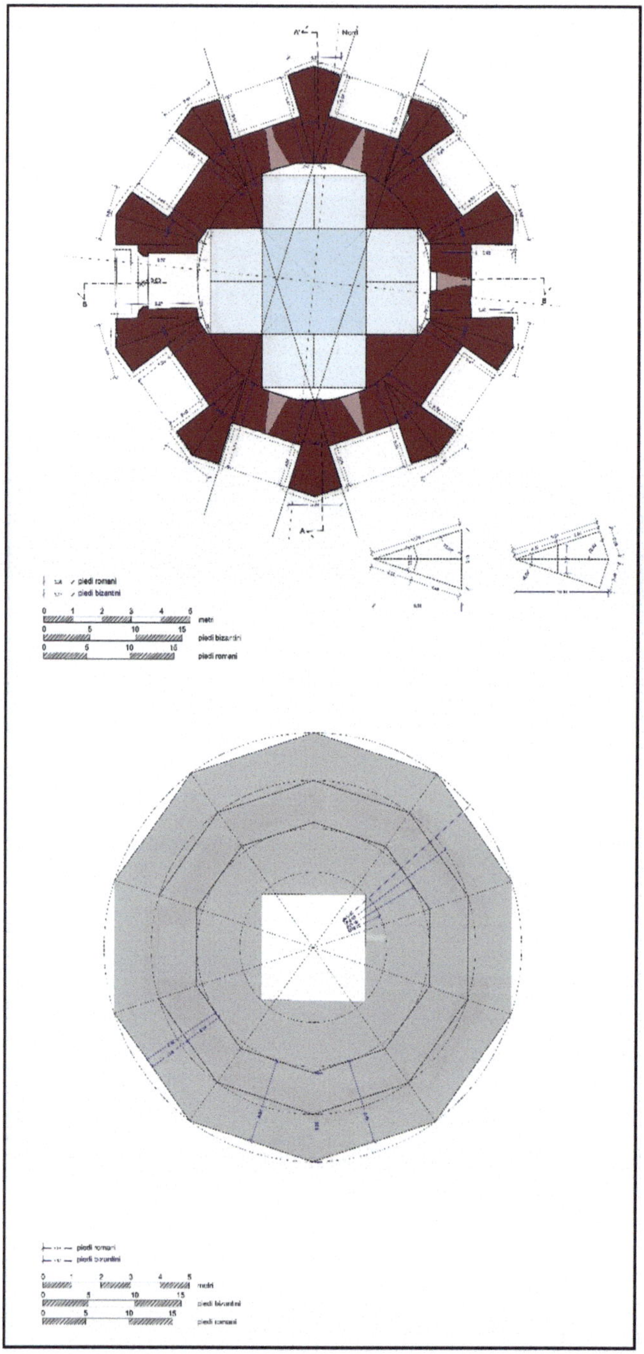

Fig. 11.4 Plan of the first floor: geometry and measurements in Byzantine feet

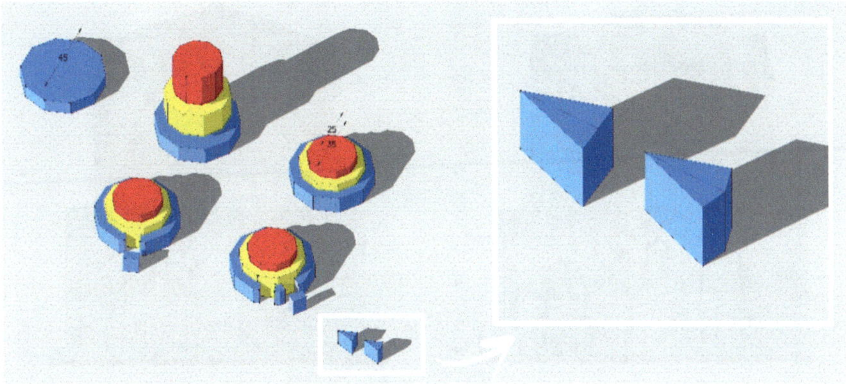

Fig. 11.5 Schematic drawing of the mausoleum volumes

An interesting geometrical quality of the decagon is that it can be divided into 10 isosceles triangles whose base angles measure 72°, and the other half of the opposite, i.e. 36°. The 10 powerful external pillars (Fig. 11.5), that are constructed on a quadrilateral made of two triangles with 10 bf hypotenuse and 9.5 bf side (angles 18°, 72°, and 90°), can be traced back to these triangles, the sum of which results in an isosceles triangle 36°, 72°, 72°, with equal sides of 10 bf and height 9.5 bf. The minimum dimension of the section of the pillar bordering the outer niches is of 3 bf (Fig. 11.4). Finally, the interior space can be easily approximated by a Greek cross, whose central square measures 11.1 bf, while the four lateral arms are rectangles with a 1/2 ratio to the square.

Regarding the upper floor of the building, it is necessary to state that the conditions of the external stone blocks do not allow, in our opinion, an accurate reading of the measurements of the existing profile. It can be hypothesised that the circumscribed circle at the base of the pilasters measured 36.65 bf, corresponding to 39 rf. The side of the inscribed dodecagon could thus be 11.33 bf, a dimension that is relatable to 12.06 rf. The upper cylinder on which the slot openings are found has an external diameter of 34 bf and an average thickness of about 2.45 bf.

The interior elevations (Fig. 11.6) are characterized by decimal measurements attributable to the unit division into 1/3, 2/3 bf. The main architecture lines of the lower floor appear to rely on a 2×3 square grid, while the higher one on a 5×8 grid (amount very close to the golden ratio). Even the arrow of the vault is attributable to the Byzantine foot and measures 6 bf. The 2×3 ratio is also present in the exterior elevation on the side of the decagon of the first level. In this case, the rectangle is displayed vertically and its measure depends on the side of the decagon of the plan. Its value is thus an irrational number derived from the measurement of the circumscribed circle of 45 bf.

Finally, it should be mentioned that other architectural elements are related to the Byzantine foot: for example, the maximum thickness of the cylinder on which the cupola rests is 4 bf. One must also highlight that certain measurements yield whole

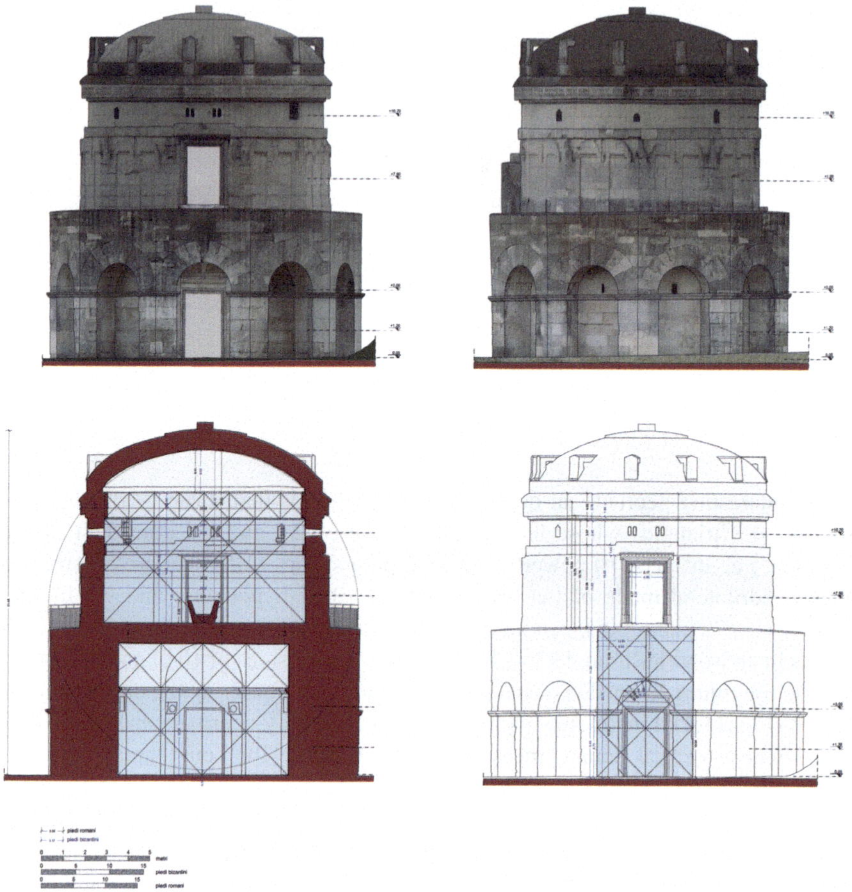

Fig. 11.6 Orthophotos; section AA' with indication of the proportions 2 × 3 and 5 × 8. 9; external elevation with indicated proportion 2 × 3

numbers in Roman feet. This is the case of the outer band decorated with a 'pincer' pattern (2 rf), the outer extent of the apse equal to 10 rf (whole number), and the pilasters of the smaller width of the gallery, which amounted to approximately 2 rf. On the ground floor the total height of the frame is 13 rf, the door height is 10 rf.

The Decagon and Boethius

In AD 526 (or according to tradition, in 524) Anicius Manlius Severinus Boethius, questor, patrician, consul and *magister officiorum* at the Theodosian court, died in Pavia, imprisoned and killed by Theodoric. The philosopher, as we know, is credited with the term *quadrivium*, a word that was used to describe the art of late-ancient scientific knowledge. The four disciplines—arithmetic, music, geometry and astronomy—have their roots in Greek tradition and constituted the 'preparatory paths of philosophy'. Architecture students had to follow such structured science, and were of course also trained in practical themes: the balance between the theoretical and the operational skills in late antiquity certainly had different outcomes in Roman society and Byzantine society (Briggs 1927; Frothingham 1909; Kostof 2000; Meek 1952; Schibille 2009; Vagnetti 1980). Within the present study, some reflections on the possible practical application (in the design phase) of the theoretical knowledge possessed by Boethius at that time certainly appear necessary.

The writings on the scientific subject attributed to Boethius have only partly reached us, unfortunately fragmented and incomplete, as attested by the relative philological studies. While the *De Institutione Arithmetica* reached us intact, the same cannot be said for other sections: *De Institutione Musica*, *De Geometria* and the Astronomy. The first work contains the knowledge of Nicomaco di Gerasa (already translated by Apuleio). The sources of the third have to be found in the documents of *Euclid's Elements*, while the astronomical works of Tolomeo were used for the fourth (see the letter between Theodoric and Boethius reported by Cassiodoro, *Variae*, I, 45, 4).

Scholars have long debated the authenticity of the two geometry books attributed to Boethius (Folkerts 1970), highlighting the incongruous traits and elements that move the dating of the earliest manuscripts to the eleventh century. However, some fragments are contained in the third and fourth book of the *Ars Geometriae et Arithmeticae* in five books (Boezio 1867). In this work, which will remain a point of reference for Cassiodorus and the measurements of the Middle Ages, a brief description of the decagon appears (book II, XXX). The short passage describes the properties of the decagon, not so much from the geometrical point of view as from the arithmetic point of view across the figurate numbers. In the book on geometry, the author associates the decagonal number 370 with the figure of the decagon (Fig. 11.7), which, although not appearing in *De Institutione Arithmetica*, can be traced back to the same arithmetic principles of the polygonal numbers (Incerti et al. 2017: 76).

It is clear that the geometric rules followed by the anonymous designer of the Mausoleum belonged not so much to the field of arithmetical calculation and the properties of particular numbers such as the decagonal ones cited in the *De Geometria*, attributed to Boethius, but to the geometric knowledge already present in the Euclid's *Elements*.

Conclusions

To conclude, an archaeoastronomical investigation has certainly yielded significant results which have extended our knowledge of this extraordinary building into topics that were previously unexplored. The critical reading of the survey measurements also allowed us to highlight the presence of a geometrical project that controlled the general

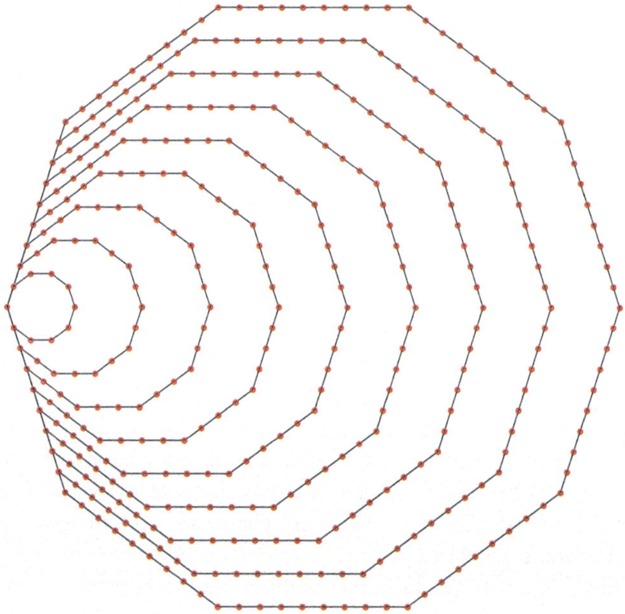

Fig. 11.7 The figure corresponding to the decagonal number 370

measures of the buildings based on Byzantine foot measurements. In addition to the encircled and circumscribed decagons, whose diameters were integer figures, other numerical relationships were found in the plans and elevations, such as: 1:2, 2:3 and 5:8. The comparison of some Roman integer measures, however, makes it clear that this second unit of measurement has also been used not so much during the project phase as during the execution of the work. Finally, we have tested the important contribution of digital models both in the phase of analysis and in the communication of complex and stratified contents such as the historical-astronomical ones.

Acknowledgements We thank the Polo Museale Regionale dell'Emilia Romagna and the Compagnia delle Misure for the use of their Faro focus3d scanner during our architectural survey of the Mausoleum. Sections "Introduction: The Foundation and the Main Topics", "A Description of the Mausoleum", "The Architectural Survey", "Archaeoastronomical Analysis", "Numbers and Geometry", "The Decagon and Boethius", "Conclusions" are by M. Incerti; Section "Survey Drawings: Methods and Procedure" by G. Lavoratti, Section "The Interactive Models for the Dissemination of the Research Project" by S. Iurilli.

References

Ambrogi, A. (1995). *Vasche di età Romana in Marmi Bianchi e Colorati*. Roma: L'Erma di Bretschneider.

Bevilacqua, F., Fabbri, R., Grillini, G. C., & Iannucci, A. (2003). Il Mausoleo di Teodorico: La pietra di Aurisina. Tecniche e strumenti di lavorazione. In F. Lenzi (Ed.), *L'Archeologia dell'Adriatico dalla Preistoria al Medioevo. Atti del Convegno Internazionale: Ravenna 7–9 Giugno 2001* (pp. 572–580). Firenze: All'Insegna del Giglio.

Bianco Fiorin, M. (1993). Il monolite del Mausoleo di Teodorico. In *Teoderico il Grande e i Goti d'Italia. Atti del XIII Congresso Internazionale di Studi sull'alto Medioevo, Milano, 2–6 Novembre 1992. Volume II* (pp. 601–610). Spoleto: Centro Italiano di Studi sull'Alto Medioevo.

Boezio, S. (1867). *De Institutione Arithmetica Libri Duo, De Institutione Musica Libri Quinque; Accedit Geometria quae Fertur Boetii e Libris Manu Scriptis Edidit Godofredus Friedlein.* Lipsiae: B.G. Teubneri.

Bovini, G. (1977). *Il Mausoleo di Teodorico.* Ravenna: Longo.

Briggs, M. (1927). *The architect in history.* Oxford: Clarendon Press.

Conti, G., & Berti, M. (1997). Lacune e integrazioni murarie nel Mausoleo di Teodorico a Ravenna. In *Lacune in Architettura: Aspetti Teorici ed Operativi. Atti del Convegno di Studi, Bressanone, 1–4 Luglio 1997* (pp. 361–378). Marghera: Edizioni Arcadia Ricerche.

De Angelis d'Ossat, G. (1962). *Studi Ravennati: Problemi di Architettura Paleocristiana.* Ravenna: Edizioni Dante.

Dyggve, E. (1957). Mausoleo di Teodorico: Le origini della cupola. In *Corso di Cultura sull'Arte Ravennate e Bizantina Università degli Studi di Bologna, Istituto di Antichità Ravennati e Bizantine, Volume 4(2)* (pp. 67–73). Ravenna: Univerity of Bologna.

Fagiolo, M. (1972). Theodericus, Christus, Sol. Nuove ipotesi sul Mausoleo. *Arheoloski Vestnik, 23*, 83–117.

Folkerts, M. (1970). *Boethius Geometrie 2.* Wiesbaden: F. Steiner.

Frothingham, A. L. (1909). The architect in history III. The architect during the Dark Ages. Parts 1, 2. *The Architectural Record, 26*, 55–70, 140–152.

Gotsmich, A. (1958). Il Mausoleo di Teodorico. In *Felix Ravenna Università degli Studi di Bologna, Dipartimento Di Archeologia, Centro di Studi per le Antichità Ravennati e Bizantine Giuseppe Bovini. Volume 3* (pp. 56–71). Ravenna: Univerity of Bologna.

Guberti, V. (1952). *Il Mausoleo di Teodorico detto anche "La Rotonda".* Ravenna: Arti Grafiche.

Haupt, A. (1913). *Das Grabmal Theoderichs des Grossen zu Ravenna.* Leipzig: Seemann.

Heidenreich, R., & Heinz, J. (1971). *Das Grabmal Theoderichs zu Ravenna.* Wiesbaden: F. Steiner.

Herz-Fischler, R. (1998). *A mathematical history of the golden number.* New York: Dover Publications.

Incerti, M., Lavoratti, G., & Iurilli, S. (2016). Survey, archaeoastronomy and communication: The Mausoleum of Theodoric in Ravenna (Italy). *Mediterranean Archaeology and Archaeometry, 4*, 437–446.

Incerti, M., Lavoratti, G., & Iurilli, S. (2017). Il Mausoleo di Teodorico: Rilievo, analisi e comunicazione. In O. Niglio & C. Visenti (Eds.), *Conoscere, Conservare, Valorizzare il Patrimonio Culturale Religioso. Volume 3: Archivi, Musei, Biblioteche* (pp. 70–78). Roma: Aracne.

Johnson, M. J. (1988). Toward a history of Theoderic's building program. *Dumbarton Oaks Papers*, 42, 73–96 (CR-Copyright © 1988 Dumbarton Oaks).

Kostof, S. (2000). The architect in the Middle Ages, East and West. In S. Kostof (Ed.), *The architect: Chapters in the history of the profession* (pp. 59–95). Berkeley: University of California Press.

Martini, A. (1883). *Manuale di Metrologia, Ossia Misure, Pesi e Monete in uso Attualmente e Anticamente Presso tutti i Popoli.* Torino: E. Loescher.

Meek, H. A. (1952). The architect and his profession in Byzantium. *Journal of the Royal Institute of British Architects, 59*, 216–220.

Messina, A. (1980). La cosiddetta "scarsella" del Mausoleo di Teodorico. In *Felix Ravenna Università Degli Studi Di Bologna, Dipartimento Di Archeologia, Centro Di Studi per Le Antichità Ravennati E Bizantine Giuseppe Bovini. Volume 1979, 117* (pp. 29–38). Ravenna: University of Bologna.

Muratori, L. A. (1738). Anonymus Valesianus. In *Rerum Italicarum Scriptores, XXIV, IV* (p. 640, X). Mediolani: Ex Typographia Societatis Palatinae.

Novara, P. (2013). Mausoleo di Teodorico. In A. Ranaldi & P. Novara (Eds.), *Restauri dei Monumenti Paleocristiani e Bizantini di Ravenna Patrimonio dell'Umanità, Parte I. Architetture* (pp. 114–116). Ravenna: Comune di Ravenna, Archidiocesi di Ravenna-Cervia, MIBAC.

Ousterhout, R. (2008). *Master builders of Byzantium*. Philadelphia: University of Pennsylvania Museum of Archaeology and Anthropology.

Piazza, V. (2009). *Mauseleo di Teodorico*. Ferrara: Scheda Soprintendenza per i Beni Architettonici e Paesaggistici per le province di Ravenna.

Piazza, V. (2013). Mausoleo di Teodorico. In A. Ranaldi & P. Novara (Eds.), *Restauri dei Monumenti Paleocristiani e Bizantini di Ravenna Patrimonio dell'Umanità* (pp. 84–86). Ravenna: Comune di Ravenna, Archidiocesi di Ravenna-Cervia, MIBAC.

Piazza, V., Bevilacqua, F., Grillini, G. C., & Pinna, D. (1998). Ravenna – Mausoleo di Teodorico. Esiti delle ultime indagini e progetto. In G. Driussi & G. Biscontin (Eds.), *Progettare i Restauri: Orientamenti e Metodi-Indagini e Materiali. Atti del Convegno di Studi, Bressanone 30 Giugno–3 Luglio 1998* (pp. 251–260). Venezia: Arcadia Ricerche.

Romano, G. (1995). *Orientamenti ad Sidera. Astronomia, riti e Calendari per la Fondazione di Templi e Città. Un esempio a Ravenna*. Ravenna: Edizioni Essegi.

Schibille, N. (2009). The profession of the architect in late antique Byzantium. *Byzantion, 79*, 360–379.

Schilbach, E. (1970). *Byzantinische Metrologie*. München: C.H. Beck.

Schilbach, E. (1991). Daktylos. In *Oxford dictionary of Byzantium* (p. 578). Oxford: Oxford University Press.

Tabarroni, G. (1973). La cupola monolitica del Mausoleo di Teodorico: Problemi tecnici e interpretativi. In *Felix Ravenna Università Degli Studi Di Bologna, Dipartimento Di Archeologia, Centro Di Studi per Le Antichità Ravennati E Bizantine Giuseppe Bovini. Volume 105–106 (4.Ser. 5/6)* (pp. 119–142). Ravenna: University of Bologna.

Underwood, P. A. (1948). Some principles of measure in the architecture of the period of Justinian. *Cahiers Archéologiques, 3*, 64–74.

Vagnetti, L. (1980). *L'Architetto nella Storia d'Occidente*. Padova: Cedam.

Chapter 12
Virtual Archaeoastronomy: Stellarium for Research and Outreach

Georg Zotti, Bernard Frischer, Florian Schaukowitsch, Michael Wimmer, and Wolfgang Neubauer

Abstract In the last few years, the open-source desktop planetarium program Stellarium has become ever more popular for research and dissemination of results in Cultural Astronomy.

In this time we (LBI ArchPro and TU Wien) have added significant capabilities for applications in Cultural Astronomy to the program, in particular a way to allow virtual 3D exploration of architecture from any period. The major part of this chapter describes our recent accomplishments for allowing its use in a multi-screen installation running both completely automated and manually controlled setups in an exhibition about Stonehenge. During the development time, also the accuracy of astronomical simulation has been greatly improved. The final part of this chapter (authored by B. Frischer) presents the latest application examples, in particular of these 3D capabilities, for Cultural Astronomy research in the Roman world.

Introduction

Communicating scientific topics in state-of-the-art exhibitions and science centres frequently involves the creation of impressive visual installations. In the 2016–2017 exhibition "STONEHENGE. A Hidden Landscape." at the MAMUZ Museum in Mistelbach, Lower Austria (MAMUZ, n.d.), LBI ArchPro presented its recent research results from the Stonehenge Hidden Landscape Project (e.g. Gaffney et al. 2012; Löcker et al. 2013). A central element of the exhibition which extended over two floors in a former machine factory hall connected with open staircases was

G. Zotti (✉) · W. Neubauer
Ludwig Boltzmann Institute for Archaeological Prospection and Virtual Archaeology (LBI ArchPro), Vienna, Austria
e-mail: Georg.Zotti@univie.ac.at; Wolfgang.Neubauer@archpro.lbg.ac.at

B. Frischer
Department of Informatics, Indiana University, Bloomington, IN, USA

F. Schaukowitsch · M. Wimmer
Institute of Visual Computing and Human-Centered Technology, TU Wien, Vienna, Austria
e-mail: florian.schaukowitsch@alumni.tuwien.ac.at; wimmer@cg.tuwien.ac.at

G. Magli et al. (eds.), *Archaeoastronomy in the Roman World*, Historical & Cultural Astronomy, https://doi.org/10.1007/978-3-319-97007-3_12

an assembly of original-sized CNC-milled replicas of most of the central trilithon horseshoe which is seen from both floors.

In the upper floor, visitors were at eye level with the lintels and could study, for example, the connections between the stones made with mortise and tenon joints, made visible by lifting one lintel. On a huge curved projection screen of 25 × 4 m size which extended along the long wall of the hall the exhibition team aimed to create an experience where the visitors could look out over the Sarsen circle into the surrounding landscape. Two possibilities were discussed:

1. A pre-rendered panoramic movie could take the visitors around the landscape. This required a storyboard quite early in the planning phase, detailed 3D modelling that might allow for animation of virtual characters or animals, weather phenomena etc. but needed lots of computing power to create about 15–20 min footage of approximately 4 FullHD frames side-by-side, and required a specialized movie player for synchronized replay over several projectors. The scenes could include everything available from the movie industry, like animations, inserts, fly-overs of the 3D landscape, and in the context of Stonehenge also the changing aspects of celestial objects, sunrise and sunset at the solstices (if properly modelled), the course of the seasons etc. However, the production effort seemed very high.
2. A programmable (scriptable) panorama viewer could be used to present artificial horizon renderings. Unfortunately, most panorama viewers were only able to display a section of a panorama in perspective projection on a regular flat screen (4:3, 16:9 etc.), and not the full width on a very wide screen (25:4 in our case) in cylindrical projection. Also, static panoramas limited the possibilities of explanations around the astronomical connections of Stonehenge to the solstices.

Given the clear astronomical peculiarities of Stonehenge and its sibling monuments, we found that the open-source desktop planetarium program Stellarium (then at version 0.13.2) could be used for our purposes, if some features for automated shows could be added and some improvements in the long-time accuracy of the program could be implemented to allow presentation of celestial motion in prehistory. Stellarium already provided various projections including cylindrical (equirectangular and Mercator) which seemed the natural choice for such a panorama wall, and scripting capabilities which allow automated shows.

Both manufacturers of dedicated high-power graphics hardware, AMD and Nvidia, offered graphics cards which could drive multi-screen setups, and an early test showed that Stellarium could work with a viewport of at least 8192 pixels width.

Over the following months, we would therefore develop Stellarium a few steps further, both in astronomical accuracy and usability as a carrier for media-rich presentations.

An Open-Source System for Virtual Archaeoastronomy

Virtual Archaeology allows the 3D visualisation of reconstructed building structures in a reconstruction of the past landscape with the help of computer graphics technologies. If the system is capable of presenting such models together with celestial objects at astronomically correct simulated positions, we can use such a system for accurate visual simulation of astronomical circumstances potentially connected to those buildings. This enables researchers to find and/or better understand astronomical orientation patterns or, if the system allows, light and shadow phenomena caused by sunlight or moonlight around such sites (Zotti 2015).

A typical sky simulation program can simulate stars and planets as they are located in the sky on any date and time and any location on Earth. For a better feeling of 'being outside in the field', a panorama photograph or artificial rendering of the landscape horizon can be added to the simulation. However, the observer usually cannot move inside the surrounding landscape, whereas archaeoastronomy often deals with extended monuments in and above the ground, and a combination of sky simulation and 3D foreground renderer had been a desire for a long time (Zotti et al. 2006).

The open-source movement has brought forth a number of high-quality programs. A highly realistic sky simulation program called Stellarium had been developed by a small group of developers around Fabien Chéreau and had already become quite mature by around 2007 (Stellarium n.d.). Over the following years it has been used by many laypeople and researchers because of its free availability, ease of use and high visual quality, although on closer investigation it was evident that it was not accurate enough for simulations in very remote times like Roman antiquity or even prehistory because some long-time effects had not been fully implemented. However, an open-source project allows own improvements and extensions, so this project seemed almost perfect to continue and extend when, during the ASTROSIM project (2008–12), we were looking for a good visualisation engine to investigate the potential astronomical orientation of Neolithic circular ditch systems in Lower Austria (Zotti and Neubauer 2015).

The Scenery3D Plugin

A first prototype of a plugin (program extension) for 3D landscape models in Stellarium had been developed in collaboration with students of the Institute of Visual Computing and Human-Centered Technology and used during the ASTROSIM project (Zotti and Neubauer 2012). At that time changes in the project structure of Stellarium did not allow the publication of this plugin. However, further large changes of the Stellarium project which now required the Qt5 framework allowed one of us (FS) to update, extend and complete the integration of the

Scenery3D plugin into a mature and efficient polygonal real-time 3D renderer that was released with Stellarium 0.13.3 in April 2015 (Zotti 2016).

This plugin allows loading a 3D model in the well-known Wavefront OBJ format (complete with MTL material and texture definition) in a way that the user is placed at eye height (configurable) over the surface of the model terrain. This model is currently rendered in a 'flat earth' approximation, i.e., on a plane tangential to the Earth's globe at the location coordinates configured in Stellarium. Actually two OBJ files can be loaded: one includes the full scene with architecture, walls, pillars etc., while a 'ground' layer without walls and pillars defines the plane where the virtual camera (i.e., the visitor's eye) will be located. Like in a computer game, the user can explore the model by moving the view over this ground layer of the model, which means that we can simply enter rooms or walk through walls without jumping onto roofs or stepping onto those walls. The plugin does not attempt to act as a complete game engine with animations, wind-moved vegetation, sound, etc. On the other hand, fine detail in the model (e.g. inscriptions on the walls of a temple) can be modelled by normal mapping, and the shadow of the Sun and Moon can be rendered diffuse by the PCSS algorithm (Fernando 2005), which should help simulating the reduced saliency of shadows cast by slim columns (see discussion of the Montecitorio Obelisk and Ara Pacis below).

The models can be presented with eyepoint (camera) coordinate values representing real-world survey coordinates e.g. in the UTM coordinate system, which allows proper investigation and reproducible results to be gained for scientific applications. We had to overcome a problem though: real-time computer graphics applications usually only work with single-precision (32-bit) floating point numbers, providing only about 7 significant decimal digits. Coordinates in metre-based UTM can easily grow to millions of metres, which means that vertex (corner) coordinates for model elements could be specified only to the nearest decimetre or even metre, which is clearly not enough. We can however split off a large 'offset' from the coordinate values of our location (either the centre coordinates of our virtual terrain or some full-metre offset for easier maintenance by humans) and work in a local coordinate system for rendering. The numerical coordinate text display simply reconstructs the original coordinates by adding the offset to the local coordinates used for rendering.

Another problem when modelling in a Cartesian survey grid like UTM: unless the location is located on the UTM zone meridian, the direction called 'Northing' does not point strictly northward, but slightly deviates because of the meridian convergence towards the poles. For a limited area typical for an archaeological site, this deviation can be compensated by a simple rotation of the model around the vertical axis.

Currently (as of version 0.15.2) the model does not deal with curvature of the Earth or terrestrial refraction, and also does not take into account that walking in the model should modify the geographic location coordinates in an order of arcseconds. Usually the modelled landscapes are small enough that these effects can be neglected. Mountains in tens of kilometres distance on the far horizon should be included as classical Stellarium 'landscape', a 2D panorama accurately rendered along the mathematical horizon.

We are convinced that such a machinery for virtual archaeoastronomy is the best tool to recreate observers' views of past landscapes, buildings and monuments under the day and night sky of corresponding past times, i.e. the 'skyscape' (Silva and Campion 2015). For best possible results, accurate data should necessarily be used. Especially for the landscape surrounding a monument of interest, LIDAR-based digital terrain models provide critically more detail than data previously available from official survey authorities. Possible users should take note that terrain models based on the freely available 90 m SRTM data are generally too coarse to achieve reliable results. As a first preview of local horizon conditions in not too steep or rough terrain, their use seems acceptable though.

Likewise, another plugin called ArchaeoLines, particularly aimed at archaeoastronomical simulations, was introduced in late 2014 (Zotti 2016). It can show diurnal arcs for the most prominent declinations (solstices, solar cross quarters, lunistices, zenith passages etc.) and azimuth indicator lines (either arbitrary azimuth or azimuth towards geographical targets like sacred mountains or places). Such simple lines can be helpful in explanations and presentations of astronomical concepts also in the context of the forthcoming exhibition. Also the landscape panoramas have been enhanced by an option to label features like mountain peaks.

Meanwhile, several deficiencies of Stellarium for celestial simulations in remote times were identified, and it was now time to improve computational accuracy for this application which should extend Stellarium's applicability if possible back into the Mesolithic. The largest impact was the replacement of the simple first-order approximation of the precessional motion of Earth's axis by a modern long-time model compatible with IAU2006 precession but applicable for ±200,000 years around the year 2000 (Vondrák et al. 2011, 2012). This model has been explicitly developed for applications like archaeoastronomy. In addition, nutation (IAU2000B from McCarthy and Luzum 2003) has been added, a small faster 'wobble' of Earth's axis in the order of several arcseconds that is however only applied ±500 years from J2000.

Stellarium's planetary positions were based on the widely used VSOP87 analytic model which has been created as fit over the numerical solution DE200 from NASA/JPL and is recommended for use in the year range −4000 to +8000. With the accurate precession model mentioned above added, we can observe in Stellarium what happens outside this time: between 8000 and 8100 the Sun moves from the 'ecliptic of date' to the ecliptic of J2000 that is VSOP's XY plane. Likewise, before −4100 the Sun (or rather Earth) is mathematically 'forced' into this plane, likely to achieve at least approximate positions outside the recommended range of dates (Fig. 12.1). This clearly indicates that for dates before −4000, VSOP87 should not be relied upon (as specified by its authors). Fortunately NASA/JPL recently has released DE431, results of a numerical simulation for the years −13,000 to +17,000 (Folkner et al. 2014). ESA has sponsored another student in the 'Summer of code in space' project 2015 so that Stellarium can now access both DE430 (high accuracy) and DE431 (long-time simulation) files to retrieve accurate planetary and lunar positions for a far longer period than before (Fig. 12.2).

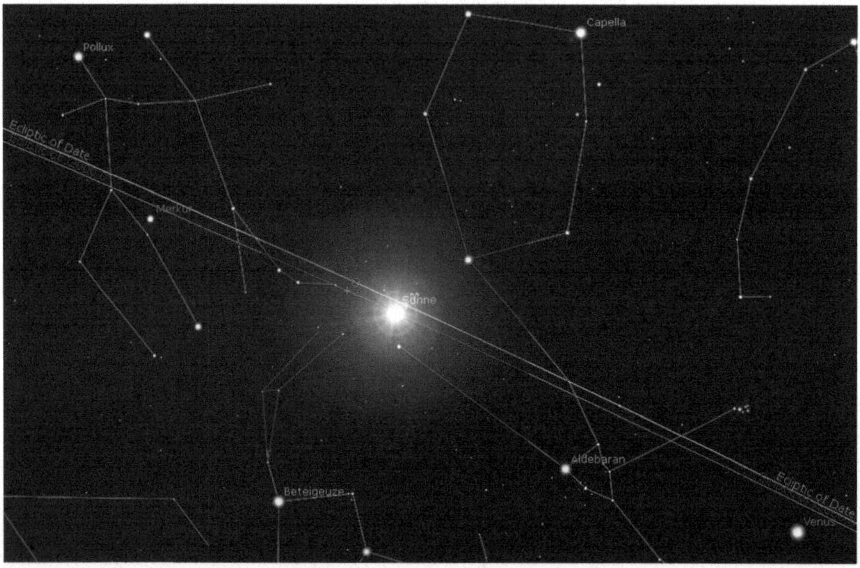

Fig. 12.1 Spring equinox, −4200, simulated in Stellarium 0.15.2 with positions from VSOP87 (outside its recommended date range). The Sun is forced to be on the J2000 ecliptical plane (lower line)

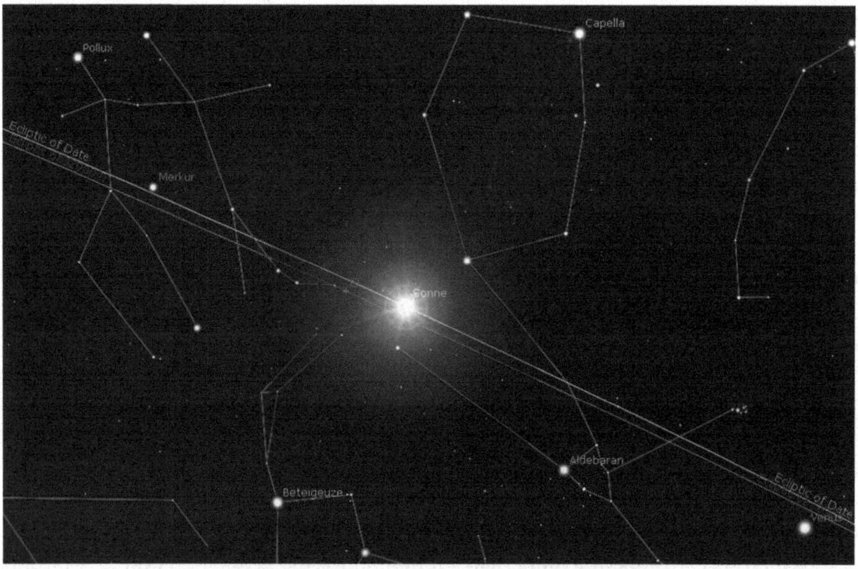

Fig. 12.2 Spring equinox, −4200, simulated in Stellarium 0.15.2 with positions from DE431. The Sun is on the ecliptic given by the new precession model from Vondrák et al. (2011, 2012). Note also the different positions of Mercury and Venus

Improvements for Astronomical Demonstrations

Stellarium has provided scripting capabilities for automatic shows since its early years. However, scripts had to be started manually. Also, Stellarium's earlier transition to the Qt5 framework had caused a temporary reduction of its multimedia capabilities. In preparation for the exhibition we have restored and extended Stellarium's multimedia capabilities so that still images and movie clips can be inserted into the screen to add visual detail to the audio narration.

The largest change for automatic shows and remote accessibility was however the development of the RemoteControl plugin. This required a 're-wiring' of the internal program modules so that their 'properties' (state variables) could be exposed and changes forwarded to other interested program modules. We developed a dedicated plugin that provides a web server interface which allows operation of Stellarium via webbrowser, where the user can switch almost every setting available in the regular program menu without showing this interface menu on the big screen of the exhibition. An alternative page layout has been added for small screens typically found on tablets. In addition, a web API allows triggering of show scripts. In the context of our exhibition, a little Raspberry Pi computer with touch screen interface has been configured to act both as master switch for the PC and 5-projector system and to send such script calls every 25 min via cronjob.

Outreach Result: The Skyscape Planetarium

Several years ago, one of the authors found that the classical immersive astronomical simulation environment of the twentieth century, the optomechanical projection planetarium, was actually not well suited when it came to the simulation of archaeoastronomical concepts and case studies of horizon astronomy: although the sky was visualized on an all-enclosing hemispherical dome, the projection included only the area above the mathematical horizon, so that the sub-horizontal context of archaeologically documented or reconstructed monuments which should have been presented in the foreground was missing. Advances in computer graphics and virtual realities promised a more applicable solution (Zotti et al. 2006).

The setup which we have now developed can best be described as 'Skyscape Planetarium' (Zotti et al. 2017), where 'skyscape' describes the inseparable combination of landscape and the sky behind, observed and experienced by human observers (Silva and Campion 2015). The projection, configured with five overlapping projectors, is limited to presenting the horizon area, but the system allows shifting the horizon up or down as required to show more sky or more of the ground, and therefore allows displaying celestial objects in a maximum altitude of about 30°, which is enough for simulating many typical views in both prehistoric and contemporary skyscapes (see Fig. 12.3).

Fig. 12.3 A Fisheye photograph of the Skyscape Planetarium in the MAMUZ Museum, combining panoramas from virtual reconstruction, astronomical simulation and explanatory inserts. On the right side, the screen is partially hidden by one of the horseshoe trilithon replicas, but the continuation of the screen adds to the immersive quality of the installation

In the context of the exhibition, we present artificial renderings of panoramic views of the landscape evolving over several millennia from the time of Mesolithic hunters to the artisans of the Bronze Age, highlighting important phases in the history of, and visiting important sites in the landscape around, Stonehenge in an automated narrated tour of about 20 min.

Most of the panoramic vistas are not related to astronomical events and Stellarium is only used to show the panoramas, often in slow pans over the landscape. We do not use the Scenery3D plugin in this exhibition, but the changing colours of simulated daylight and landscape brightness depending on solar altitude changing over time both help to create an almost 3D feeling of the landscape and standing Sarsen and Bluestone circles. Currently, Stellarium's sky does not include meteorological phenomena like clouds, but some have been included as static renderings in some panoramas when they only serve decorative purpose and are not required to move with the winds. Of course, views like the iconic summer solstice sunrise over the Heel Stone can be presented in all its glory, using high-quality panorama renderings of virtual reconstructions of Stonehenge and related sites in several phases of construction in their reconstructed landscapes.

In contrast to presenting only a pre-produced panoramic movie, using Stellarium offers the benefit of providing a complete and flexible astronomical simulation environment. This allows us, for example, to experience and demonstrate the slow shift of the solstice rising points due to changes in ecliptic obliquity. On special occasions, the regular show can be switched off, and an operator can control the application using a web browser interface on a tablet computer provided by the new RemoteControl plugin which has been developed in the context of our exhibition project.

Fig. 12.4 A scene from a 'flashlight tour' for children with the Skyscape Planetarium providing the nocturnal atmosphere

The experience of a starry virtual nightscape that fills the field of view on a big wall, first seen a few days before opening the exhibition on 20 March 2016, was stunning. Unfortunately only a few visitors are given this experience. The 20-min narration shown during daytime opening hours only shows brief moments of sunrise, sunset or twilight but does not include a nightly setting. In fact, showcase illumination during the day would brighten up any night scene to a point where the visual atmosphere would be severely degraded. However, on a few evenings, the museum offers 'flashlight tours' especially targeting primary school children. With all regular lighting switched off, the children's flashlights dancing around the replica Sarsens of the horseshoe, the projected view out of Stonehenge with the stars twinkling through the Sarsen circle panorama created from the virtual reconstruction that shows Stonehenge shortly after completion provide a stunning night atmosphere (Fig. 12.4). Seen from the centre of the room, the rightmost part of the screen is partially hidden by a trilithon replica, which means we cannot present important detail on this part of the wall. But the fact that there *is* a screen surface behind this stone definitely helps to increase the immersive experience. More sites of archaeoastronomical interest would deserve a presentation in such an installation.

Two Applications of Stellarium to Roman Archaeoastronomy

The utility of Stellarium in the field of Roman archaeoastronomy can be exemplified by two recent projects: the identification of the sanctuary at Hadrian's Villa conventionally called the Antinoeion; and an assessment of the validity of an influential theory about the relationship of the Montecitorio Obelisk and the Ara Pacis Augustae in Rome. Detailed reports about both projects have been written (see Frischer 2017–2018; Frischer et al. 2016, 2017), so we can be brief here.

The Antinoeion

Antinous was a young man from Bithynia who became Hadrian's lover in the 120s AD (Galimberti 2012). In October of 130, he drowned in the Nile while on a trip to Egypt with the Emperor. Hadrian was devastated by this loss and bestowed cult honors on his friend, founding the city of Antinoopolis in Egypt and building temples all over the Empire (Galli 2012).

A feature conventionally called the Antinoeion is located at Hadrian's Villa in Tivoli. Discovered in 1998, the site was excavated from 2002 to 2004 under the direction of Zaccaria Mari (Mari 2012; Mari and Sgalambro 2007). The area measures ca. 63 × 50 m located adjacent to the double road leading to the so-called Vestibule. The perimeter was surrounded by a wall, within which were discovered two identical temples, ca. 15 × 9 m in size: one to the northwest, the other to the southeast. They faced each other across a plaza into which were cut water channels and planters. Springing off the western side of the plaza is a large colonnaded exedra, in the middle of which was another temple. For the sake of convenience, we refer to the three shrines as Temples 1, 2, and 3 (Fig. 12.5).

Archaeological investigation of the site was motivated from the first to prove the theory that somewhere at Hadrian's Villa there must have been a sanctuary dedicated to Antinous. The first to propose the theory was Kähler (1975: 42), who argued that the likely spot was the Canopus, but, despite various efforts (e.g. see Hannestad 1982), no evidence was ever found there to support Kähler's theory.

In the early 2000s, Mari's excavations of the site now conventionally called the Antinoeion turned up dating material placing it securely in the period after Antinous' death in AD 130. They also uncovered many fragments of Egyptianizing sculpture (Mari 2012), decorative elements clearly appropriate for the cult of a young man who drowned in the Nile and who was sometimes assimilated into his new cult to Osiris. Mari also argued that the obelisk now in the Villa Borghese on the Pincian Hill in Rome came from the Antinoeion. Furthermore, Mari produced documentary evidence that the Egyptianizing statues in the Gregorian Egyptian collection of the Vatican Museums were also found on land adjacent to the Antinoeion and thus, according to Mari, must have originally been erected in the sanctuary.

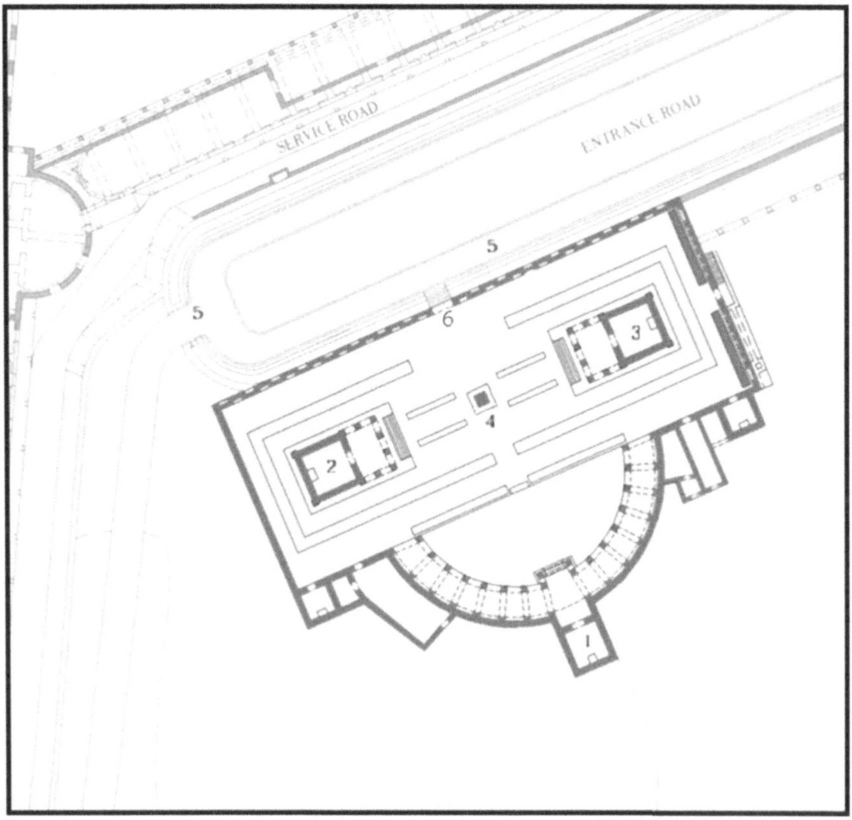

Fig. 12.5 The Antinoeion at Hadrian's Villa. Within the sanctuary are three temples (1, 2, 3), a platform which may have supported an obelisk (4), the main entrance road to the Villa (5), and the East Portal (6), leading from the road into the sanctuary. East is at top (courtesy: Matthew R. Brennan, Virtual World Heritage Laboratory, Indiana University)

Mari's work in bringing to light the Antinoeion appeared to have brought the search for Antinous' monument at the Villa, initiated by Kähler, to a successful conclusion. But in the 10 year period after Mari concluded his fieldwork, skepticism about some of his conclusions was expressed, notably in Grenier (2008). Our virtual archaeoastronomical project had the goal of trying to see if there were any alignments between celestial features and the built features of the sanctuary and, if so, to gauge their impact on the validity of Mari's interpretation of the site as an Antinoeion.

We took as our point of departure the fact that the sanctuary was clearly Egyptian in inspiration, décor, and cult. We therefore used as our working assumption the conclusions about the astronomical orientation of Egyptian temples reached by Magli (2013) and by the Egyptian-Spanish Mission on Egyptian Archaeoastronomy (henceforth ESMEA; see Belmonte et al. 2009).

Magli showed the importance of the concept of the 'sacred landscape' in the layout and design of Egyptian monuments such as pyramids and temples. ESMEA investigated whether Egyptian temples tended to be oriented to the Nile, as is commonly believed, or to appropriate astronomical phenomena. ESMEA surveyed a remarkable 98% of existing shrines and temples in Egypt from all periods found at nearly 400 sites around the country. The mission found that temples near the Nile were oriented to it but in such a way that astronomical orientations were also common. Both determinants of temple orientation could operate in tandem because it was possible to choose a site near the Nile, which does not run perfectly straight, such that the siting would point the temple to the correct orientation in the sky.

Applying these findings to the Antinoeion, we might equate to the Nile (in purely functional—not symbolic—terms) the double road, along which the sanctuary is located (no. 5 on Fig. 12.5). It was an obvious determinant of the orientation of the complex, though nothing required siting the sanctuary here—there was ample available land elsewhere at Hadrian's Villa. In this connection, we should also note that within the perimeter walls of the sanctuary the built features could be positioned and oriented with complete freedom. So, it makes sense to consider whether we find the same complementary terrestrial and astronomical factors dictating the orientation of the Antinoeion as ESMEA discovered in its Egyptian study.

There were three temples within the complex, each with its own orientation (nos. 1, 2, and 3 on Fig. 12.5). The possible astronomical orientation of each of these needs to be considered separately. Owing to space limitations, we concentrate here on Temple 1. The overall orientation of the sanctuary, from the point of view of the ancient visitor entering from the access road, runs from the East Portal (no. 6 on Fig. 12.5) to Temple 1. This was clearly the important axis for the ancient visitor to the site.

We created a 3D digital model to reflect the Mari-Sgalambro reconstruction of the sanctuary (Mari and Sgalambro 2007: 85, Fig. 3) set into the landscape reconstructed using data from the survey of Hadrian's Villa by the Scuola degli Ingegneri di Roma in Lanciani 1906 as augmented by the terrain map of the Tivoli area published at a scale of 1:25,000 by the Istituto Geografico Militare. The accuracy of the latter for the cardinal points was graciously confirmed for us by Valerio Baiocchi, an expert on the cartography of central Italy. Putting the digital model into Stellarium and setting the time for the years in the 130s AD after the death of Antinous, we were able to take observations of the day and night-time sky throughout the year, looking for possible celestial alignments of interest. We discovered that from the vantage point of someone looking east and standing anywhere along the axis running from the East Portal to Temple 1, the Sun would have been visible rising behind the mountains to the east on the summer solstice in the 130s (Fig. 12.6). Because of the narrowness of the East Portal, the view of sunrise would have been blocked most of the other days of the year.

Once we had discovered the sanctuary's orientation toward sunrise on the summer solstice, we interpreted it by noting that there was a major Roman holiday on the summer solstice, the festival of Fors Fortuna (Degrassi 1963: 472–473). This festival was relevant to an Egyptianizing sanctuary because we know that the Romans assimilated Fors Fortuna to Isis (Champeaux 1982: 210).

Temple 1 is located at the end of the axis of entrance opposite the East Portal. We know that when Antinous died, the grief-stricken Emperor carved out a new constellation in his memory. As Ptolemy (Almagest VII.5) reports, it was located in what had formerly been the lower part of Aquila. The idea was that Antinous is a

Fig. 12.6 A Stellarium simulation showing the Sun rising at 5: 15 am CET on the summer solstice (24 June) in the year AD 135 as seen by someone looking eastward toward the East Portal of the Antinoeion while standing on the axial line running from the East Portal to Temple 1

Ganymede figure, snatched to the heavens by Zeus, symbolized by the eagle. When we stood along the axis of entry and looked west in the direction of Temple 1 all year during the mid-130s AD, we discovered that the stars making up Antinous set on axis behind Temple 1 and that their heliacal settings occurred during a two-week period around 27 November, the date of Antinous' birthday (Fig. 12.7). Mari had already associated Temple 1 with Antinous on the level of iconography: he had noted that the Osirantinous telamones, originally known to be from Tivoli and presently in the Sala Greca of the Vatican Museum, would fit onto the porch of the temple. Mari also suggested that one of the several Osirantinous statues found out of context at the Villa might have been the cult statue. Our finding that Antinous' constellation started to set on axis with Temple 1 and had its heliacal setting around the time of Antinous' birthday added a new reason to support Mari's identification of the shrine as an Antinoeion.

The Relationship Between the Ara Pacis Augustae and the Montecitorio Obelisk

The second project in which Stellarium was used involved one of the first obelisks brought to Rome—the Montecitorio obelisk brought by Augustus from Heliopolis—and one of the best-preserved Augustan monuments, the Ara Pacis Augustae. The so-called Monte Citorio obelisk was brought by Augustus to Rome in 10 BC and set up in the Campus Martius about 90 m west of the Ara Pacis, which had been under construction since 13 BC and was completed in 9 BC. Explicitly dedicated by Augustus to the Sun

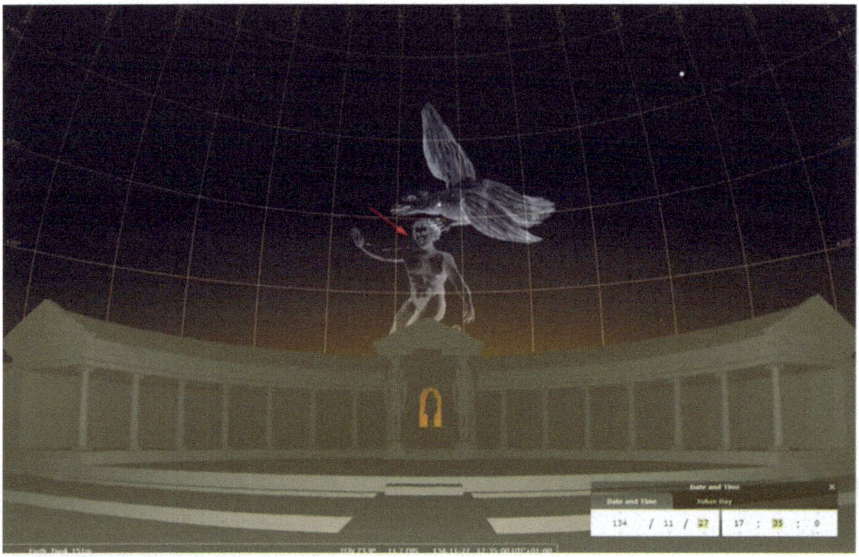

Fig. 12.7 A Stellarium simulation showing the constellation Antinous (red arrow) seen by someone standing on the axial line from the East Portal to Temple 1 of the Antinoeion at Hadrian's Villa. The constellation (taken from J.E. Bode's *Uranographia* (1801) appears just before its heliacal setting on axis behind Temple 1 at 6:35 pm CET on AD 27 November 134. 27 November was Antinous' birthday

god, as the still extant inscription attests, this obelisk was used to support a sphere that served as the gnomon for a monumental timekeeping device, something we know from a famous chapter in book 36 of Pliny's *Natural History*.

Since the sixteenth century, scholars have debated the nature of the timepiece, alternating between a meridian and a huge horizontal sundial. Buchner (1976) made a case in favor of the sundial theory. Buchner worked out a detailed reconstruction of its design and dimensions as part of a more general theory to explain the spatial and ideological relationship between the obelisk and the Ara Pacis. The theory had three parts: first, the set-up of the monuments, that is, exactly where they were sited in the ancient city. This is not obvious since, although we do know where the Ara Pacis stood, we do not know precisely where the obelisk was erected. All we know is that it was somewhere under the modern building at Piazza del Parlamento 3. Moreover, Buchner speculated that there was a vast pavement, 70 × 160 m, on which was inscribed a horizontal sundial. He thought the sphere we know from Pliny that stood atop the obelisk was the gnomon for the sundial. As noted, other scholars since the eighteenth century thought that there was no such pavement and no sundial but just a meridian line, that is, a single line running due north of the obelisk. This matter is still under debate and recently opinion has shifted back to the idea that the timepiece was simply a meridian, but clearly only new fieldwork (presently underway by a team led by Bernard Frischer) will resolve this issue.

The second element concerns the environmental effect caused by the set-up. Buchner thought that the effect concerned the shadow cast by the obelisk on

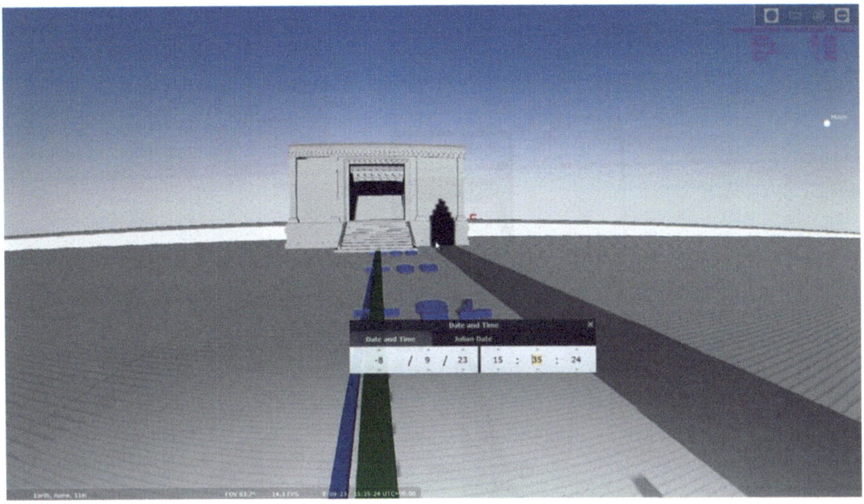

Fig. 12.8 A Stellarium simulation showing the shadow of the Montecitorio Obelisk seen by someone standing on the axial line running from the Ara Pacis Augustae to the Montecitorio Obelisk. The shadow misses entering into the entrance into the altar, which it strikes at 4:35 pm CET on 23 September 9 BC (Augustus' birthday)

Augustus' birthday, 23 September. In particular, he speculated that the shadow went down the equinoctial line inscribed on the pavement right into the middle of the Ara Pacis. Finally, the third element was the message implicitly transmitted by the effect. Buchner thought it was that Augustus was *natus ad pacem*, or 'born to bring peace'.

The purpose of our study was to test the validity of Buchner's theory about the relationship between the obelisk and Ara Pacis (for details, see Frischer et al. 2017). In assessing the validity of the theory, we started by commissioning two independent surveys of all the monuments in question, and these determined that Buchner had made a major error in siting the obelisk: he had it 4 m too far east and 2 m too far north. Once this error was corrected, we made a computer simulation of the area and used Stellarium for the year 9 BC to see whether the obelisk's shadow really proceeded along the equinoctial line into the middle of the altar on Augustus' birthday. The theory turned out to be wrong: the shadow did not enter into the western doorway of the altar but hit the west façade off axis to the south (Fig. 12.8).

Once Buchner's theory had been critiqued, Stellarium was used to support new empirical observations of the virtual Campus Martius. We found that on 48 days of the year, the shadow of the obelisk did enter into the Ara Pacis Augustae (Fig. 12.9), so Buchner may well have been on the right track but erred about the specific date(s). Moreover, virtual observations gradually produced a new insight: while the shadow cast by the obelisk toward the Ara Pacis Augustae was certainly one environmental effect created by the spatial relationship between the obelisk and altar, two other effects were equally relevant: the appearance of the solar disk seemingly centered on the top of the obelisk, as seen by an ancient observer standing along the axial line linking the altar and the obelisk (Fig. 12.10); and the rising Sun seen on axis with the altar by an observer standing on the axial line with the obelisk to his back (Fig. 12.11).

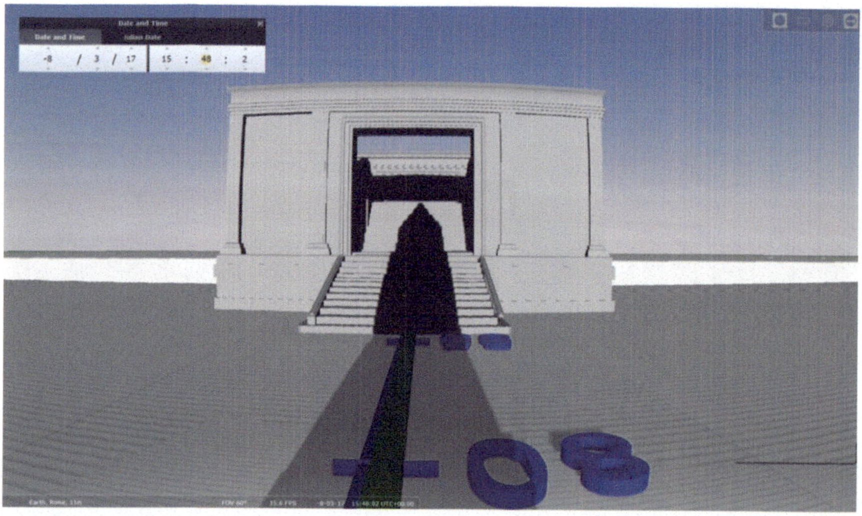

Fig. 12.9 A Stellarium simulation showing the shadow of the Montecitorio Obelisk entering into the middle of the Ara Pacis Augustae at 4:48 pm CET on 17 March 9 BC. Using Stellarium, a total of 48 days of the year were found when the shadow entered the interior of the monument. Contrary to what Buchner's theory required, Augustus' birthday (23 September) was not one of those dates

Fig. 12.10 A Stellarium simulation showing the solar disk centered on the top of the Montecitorio Obelisk as seen by someone looking west while standing on the axial line between the Ara Pacis Augustae and the obelisk at 4:33 pm CET on 23 March 9 BC. Such alignments were found on a total of 224 days of the year

Fig. 12.11 A Stellarium simulation showing the Sun rising on axis behind the Ara Pacis Augustae as seen by someone standing on the axial line running from the Montecitorio Obelisk to the altar at 4:48 am on 22 May 9 BC. Such alignments were found on a total of 111 days of the year

Frischer et al. (2017) presents the evidence as well as a series of interpretations linking these solar effects to Augustus' dedication of the obelisk to the Sun god (*CIL* 6.702 = *ILS* 91).

Acknowledgments The Skyscape Planetarium was developed as part of the exhibition project 'STONEHENGE. A Hidden Landscape' at the MAMUZ Museum Mistelbach in Austria, which provided considerable developing time for the new and improved features presented here that were published in Stellarium version 0.15.0 (released 31 July 2016).

The 3D models and renderings of the Stonehenge landscape were created by LBI ArchPro's partner 7reasons.

In the weeks before the opening, we were supported also by Stellarium maintainer Alexander Wolf (Barnaul, Russia) who even created a new customized installer package with some critical corrections built overnight just in time to set up at the Museum 2 days before the opening.

Florian's work on the RemoteControl plugin was supported by the ESA Summer of Code in Space programme 2015.

The Ludwig Boltzmann Institute for Archaeological Prospection and Virtual Archaeology (http://archpro.lbg.ac.at) is based on an international cooperation of the Ludwig Boltzmann Gesellschaft (Austria), the University of Vienna (Austria), the Vienna University of Technology (Austria), ZAMG-the Austrian Central Institute for Meteorology and Geodynamics (Austria), the Province of Lower Austria (Austria), Airborne Technologies (Austria), 7reasons (Austria), the Austrian Academy of Sciences (Austria), the Austrian Archaeological Institute (Austria), RGZM-the Roman-Germanic Central Museum Mainz (Germany), the National Historical Museums—Contract Archaeology Service (Sweden), the University of Birmingham (England), the Vestfold County Council (Norway) and NIKU-the Norwegian Institute for Cultural Heritage Research (Norway).

References

Belmonte, J. A., Shaltout, M., & Fekri, M. (2009). Astronomy, landscape, and symbolism: A study of the orientation of ancient Egyptian temples. In J. A. Belmonte & M. Shaltout (Eds.), *Search of cosmic order. Selected essays on Egyptian archaeoastronomy* (pp. 213–283). Cairo: Supreme Council of Antiquities Press.

Buchner, E. (1976). Solarium Augusti und Ara Pacis. *Römische Mitteilungen, 83*, 19–65.

Champeaux, J. (1982). *Fortuna. Le Culte de la Fortune à Rome et dans le Monde Romain. 2 Volumes*. Rome: Publications de l'École Française de Rome.

Degrassi, A. (1963). *Fasti et Elogia, Inscriptiones Italiae, Volume 13, Fasc. 2*. Rome: Istituto Poligrafico dello Stato.

Fernando, R. (2005). Percentage-closer soft shadows. In *SIGGRAPH '05: ACM SIGGRAPH 2005 Sketches*. New York: ACM Press.

Folkner, W. M., Williams, J. G., Boggs, D. H., Park, R. S., & Kuchynka, P. (2014). *The planetary and lunar ephemerides DE430 and DE431*. JPL/NASA (IPN Progress Report 42–196). https://ipnpr.jpl.nasa.gov/progress_report/42-196/196C.pdf

Frischer, B. (2017–2018). Edmund Buchner's Solarium Augusti: New observations and simpirical studies, with technical appendices by Paolo Albèri Auber, David Dearborn, John Fillwalk, Mika Kajava, and Stefano Floris. *Rendiconti della Pontificia Accademia Romana di Archeologia, 89*, 3–100.

Frischer, B., Zotti, G., Mari, Z., & Vittozzi, G. C. (2016). Archaeoastronomical experiments supported by virtual simulation environments: Celestial alignments in the Antinoeion at Hadrian's Villa (Tivoli, Italy). *Digital Applications in Archaeology and Cultural Heritage, 3*, 55–79.

Frischer, B., Pollini, J., Cipolla, N., Capriotti, G., Murray, J., Swetnam-Burland, M., Galinsky, K., Häuber, C., Miller, J., Salzman, M. R., Fillwalk, J., & Brennan, M. R. (2017). New light on the relationship between the Montecitorio Obelisk and the Ara Pacis of Augustus. *Studies in Digital Heritage, 1*, 18–119.

Gaffney, C., Gaffney, V., Neubauer, W., Baldwin, E., Chapman, H., Garwood, P., Moulden, H., Sparrow, T., Bates, C. R., Löcker, K., Hinterleitner, A., Trinks, I., Nau, E., Zitz, T., Floery, S., Verhoeven, G., & Doneus, M. (2012). The Stonehenge hidden landscapes project. *Archaeological Prospection, 19*, 147–155.

Galimberti, A. (2012). Adriano e Antinoo nelle fonti storiche. In M. S. Ragni (Ed.), *Antinoo. Il Fascino della Bellezza* (pp. 30–37). Milan: Electa.

Galli, M. (2012). Il culto e le immagini di Antinoo. In M. S. Ragni (Ed.), *Antinoo. Il Fascino della Bellezza* (pp. 38–63). Milan: Electa.

Grenier, J.-C. (2008). *L'Osiris Antinoos, Cahiers Égypte Nilotique et Méditerranéenne 1*. Montpellier.

Hannestad, N. (1982). Über das Grabmal des Antinoos. Topographische und thematische Studien im Canopus-Gebiet der Villa Adriana. *Analecta Romana Instituti Danici, 11*, 69–108.

Kähler, H. (1975). Zur Herkunft des Antinousobelisken. *Acta ad Archaeologiam et Artium Historiam Pertinentia, 6*, 35–44.

Lanciani, R. (1906). *La Villa Adriana. Guida e Descrizione*. Rome: Tip. della R. Accademia dei Lincei.

Löcker, K., Baldwin, E., Neubauer, W., Gaffney, V., Gaffney, C., Hinterleitner, A., Garwood, P., Trinks, I., & Wallner, M. (2013). The Stonehenge Hidden Landscape Project – Data acquisition, processing, interpretation. In W. Neubauer et al. (Eds.), *Archaeological prospection. Proceedings of the 10th International Conference on Archaeological Prospection, Vienna May 29th–June 2nd, 2013* (pp. 107–109). Vienna: Austrian Academy of Sciences Press.

Magli, G. (2013). *Architecture, astronomy and sacred landscape in ancient Egypt*. Cambridge: Cambridge University Press.

MAMUZ Exhibition website: http://www.mamuz.at/en/the-museum/museum-mistelbach/ausstellungen-museum-mistelbach/4-stonehenge-a-hidden-landscape?set_language=en (seen on 27 June 2017).

Mari, Z. (2012). Antinoo a Villa Adriana. In M. S. Ragni (Ed.), *Antinoo. Il Fascino della Bellezza* (pp. 79–91). Milan: Electa.
Mari, Z., & Sgalambro, S. (2007). The Antinoeion of Hadrian's Villa: Interpretation and architectural reconstruction. *American Journal of Archaeology, 111*, 83–104.
McCarthy, D. D., & Luzum, B. J. (2003). An abridged model of the precession-nutation of the celestial pole. *Celestial Mechanics and Dynamical Astronomy, 85*, 37–49.
Silva, F., & Campion, N. (Eds.). (2015). *Skyscapes: The role and importance of the sky in archaeoastronomy*. Oxford: Oxbow Books.
Stellarium website: http://stellarium.org (seen on 21 June 2017).
Vondrák, J., Capitaine, N., & Wallace, P. (2011). New precession expressions, valid for long time intervals. *Astronomy & Astrophysics*, 534, A22 (19 pages).
Vondrák, J., Capitaine, N., & Wallace, P. (2012). New precession expressions, valid for long time intervals (Corrigendum). *Astronomy & Astrophysics*, 541, C1 (1 page).
Zotti, G. (2015). Visualization tools and techniques. In C. L. N. Ruggles (Ed.), *Handbook for archaeoastronomy and ethnoastronomy, Volume 1* (pp. 445–457). New York: Springer.
Zotti, G. (2016). Open-source virtual archaeoastronomy. *Mediterranean Archaeology and Archaeometry, 16*(4), 17–23.
Zotti, G., & Neubauer, W. (2012). A virtual reconstruction approach for archaeoastronomical research. In G. Guidi & A. C. Addison (Eds.), *Proceedings of the virtual systems in the information society* (pp. 33–40). Milano: IEEE.
Zotti, G., & Neubauer, W. (2015). Astronomical and topographical orientation of Kreisgrabenanlagen in Lower Austria. In F. Pimenta, N. Ribeiro, F. Silva, N. Campion, A. Joaquinito, & L. Tirapicos (Eds.), *SEAC2011: Stars and stones: Voyages in archaeoastronomy and cultural astronomy* (pp. 188–193). Oxford: Archaeopress.
Zotti, G., Wilkie, A., & Purgathofer, W. (2006). Using virtual reconstructions in a planetarium for demonstrations in archaeo-astronomy. In C. S. Lanyi (Ed.), *Third Central European multimedia and virtual reality conference* (pp. 43–51). Veszprém: Pannonian University Press.
Zotti, G., Schaukowitsch, F., & Wimmer, M. (2017). The Skyscape planetarium. In The marriage of astronomy and culture: Theory and method in the study of cultural astronomy (Papers from the 2016 SEAC Conference). *Culture and Cosmos*, Vol. 21 no. 1–2.

The manufacturer's authorised representative in the EU is Springer Nature Customer Service Centre GmbH, Europaplatz 3, 69115 Heidelberg, Germany. If you have any concerns regarding our products, please contact ProductSafety@springernature.com

Printed and bound by CPI Group (UK) Ltd, Croydon, CR0 4YY
22/09/2026
02221063-0011